Routledge Revivals

Reform as Reorganization

As the fourth report in a series on the Governance of Metropolitan Areas, *Reform as Reorganization* explores the welfare and development of metropolitan America in terms of political reorganization. Originally published in 1974, this study reflects on metropolitan problems and governmental structure to provide some new options for policy makers and an overview of what political action can be taken. This title will be of interest to students of Environmental Studies as well as professionals.

Reform as Reorganization

Royce Hanson, Julius Margolis, Melvin R. Levin and William Letwin

RFF PRESS
RESOURCES FOR THE FUTURE

First published in 1974
by Resources for the Future, Inc.

This edition first published in 2016 by Routledge
2 Park Square, Milton Park, Abingdon, Oxon, OX14 4RN

and by Routledge
711 Third Avenue, New York, NY 10017

Routledge is an imprint of the Taylor & Francis Group, an informa business

© 1974 Resources for the Future, Inc.

Publisher's Note
The publisher has gone to great lengths to ensure the quality of this reprint but points out that some imperfections in the original copies may be apparent.

Disclaimer
The publisher has made every effort to trace copyright holders and welcomes correspondence from those they have been unable to contact.

A Library of Congress record exists under LC control number: 73019348

ISBN 13: 978-1-138-95644-5 (hbk)
ISBN 13: 978-1-315-66571-9 (ebk)
ISBN 13: 978-1-138-95649-0 (pbk)

Reform as Reorganization

Reform as Reorganization

Papers by

ROYCE HANSON

JULIUS MARGOLIS

MELVIN R. LEVIN

WILLIAM LETWIN

Published by Resources for the Future, Inc.

Resources for the Future is a nonprofit corporation for research and education in the
development, conservation, and use of natural resources and the improvement of the
quality of the environment. It was established in 1952 with the cooperation of the Ford
Foundation. Part of the work of Resources for the Future is carried out by its resident
staff; part is supported by grants to universities and other nonprofit organizations.
Unless otherwise stated, interpretations and conclusions in RFF publications are those of
the authors; the organization takes responsibility for the selection of significant subjects
for study, the competence of the researchers, and their freedom of inquiry.

This study is the fourth in a series of papers resulting from an RFF-sponsored project
conducted by an informal Commission on the Governance of Metropolitan Regions,
chaired by Charles M. Haar of the Harvard Law School. Lowdon Wingo, formerly director of RFF's program of regional and urban studies, is chairman of the department of
city planning at the University of Pennsylvania. The manuscript was edited by Margaret
Ingram.

RFF editors: Mark Reinsberg, Joan R. Tron, Ruth B. Haas, Margaret Ingram

Contents

Commission on the Governance of Metropolitan Regions

Foreword

The papers in this volume and its companions are products of a long-standing interest of Resources for the Future in the welfare and development of metropolitan America. More particularly, they stem from an RFF-sponsored project that was launched in the spring of 1970 with the convening in Washington of an informal Commission on the Governance of Metropolitan Regions. Chaired by Charles M. Haar of the Harvard Law School, it is composed of scholars, practitioners, and experienced observers of the metropolitan scene.

"Governance" in the title—The Governance of Metropolitan Regions—is meant to imply something more than government. Webster defines it as "conduct, management, or behavior; manner of life" in addition to "method or system of government or regulation." In these papers, and those that will follow, the authors are concerned not only with the apparatus and process of government in the ordinary sense but also with the total interaction among people in their public capacities and interests, and between people and the public institutions. The dominating question is: How can the governance of metropolis be improved? And next: What must we learn to achieve this? Unless early progress is made in these directions the danger that hard-pressed American cities will crack under the multiple strains of old and new problems will be very real.

RFF did not embark on this effort expecting that metropolitan political reorganization would solve all metropolitan problems, but we are inclined to think that it will help, because we have seen so many obvious steps frustrated by the way in which history has organized our urban political life. Although the reform of the institutions of metropolitan governance is hardly a sufficient condition for the solution of major urban problems, our intuition is strong that it is a necessary one.

This, then, is the theme of the RFF project—the interrelationship of metropolitan problems and governmental structure. It has formed the basis of the deliberations of the Commission and has guided the preparation of the exploratory papers published in this series. The papers do not exhaust the issues of metropolitanism; their purpose is to add some dimensions to an already rich literature, some options for the policy makers. No blueprint for the future is presented, no definitive list of recommendations; the results hoped for are breadth of view, depth of perception at several critical places, and illumination of practical alternatives for action.

Many people contributed to this effort. Charles M. Haar, as chairman, negotiated the contribution of papers and materials. Lowdon Wingo first proposed the Commission as an effective exploratory device, administered the program for RFF, and oversaw the publication of these volumes. Daniel Wm. Fessler of the University of California Law School at Davis has been a valued advisor throughout. Professors Daniel M. Holland of M.I.T. and Karl Deutsch of Harvard made valuable contributions from their time and experience. Michael F. Brewer, while Vice President of RFF, was a faithful and effective participant in the enterprise from its inception. Add to these almost twenty authors and an equal number of Commission members and you have the elements for some new insights into the metropolitan problem. Some are to be found in these papers; more are to be hoped for from succeeding phases of the project.

Joseph L. Fisher
President, Resources for the Future, Inc.

Introduction: The Ideal and the Real in the Reform of Metropolitan Governance

LOWDON WINGO

The reform of political institutions is an exercise with sometimes overlooked Platonic characteristics. Governance is the ever-changing and imperfect shadow of ideal forms. Political philosophers from Plato to Marx have sought to apprehend and describe some higher, more universal, purer form of human relations—of ethical politics, if you will—by which the interests of individual men and their constituent groups could be made coherent, for the specification of these relations is the base on which the nature of government and politics must rest.

If society is viewed as a diffuse organism whose ultimate responsibility is for the whole to grow and prosper, as in the case of the termite colony or beehive, then each man is a cell analogue, and governance the process by which each is identified with an essential role and made to carry out the tasks germane thereto. It is the homeostatic mechanism by which the social organism is made responsive to the incremental changes in a particular environment, or simply the program embedded in each constituent element, which, together with external circumstances, defines its activities. In this model of man and society, governance is *apolitical*; the elements have no interest in the distribution of power or of the benefits of organismic success—technique is all.

Personality alone makes governance a political problem. Personality stems from the awareness of self, of one's utter uniqueness and separateness from all others. But personality has needs that go beyond physiological sustenance and the mobilization of energies around a task, needs that can only be satisfied by one's relationship with others; the awareness of separateness is thus compromised by the reality of dependence. Assessing recent British exploration of the reform of subnational governmental structures, William Letwin's article in this collection brings this psychological base of government to the fore: the freedom of the individual to his own kind of self-realization supports the entire logic of representative government and political participation. Such a relationship calls for a form of governance that goes beyond servo-mechanical properties, that plays, in effect, a mediatory role between the individual's personality demands and the larger group. Power becomes important at the micro level because it influences the execution of this role; the benefits *of* society come to mean the benefits *to be distributed by* society, and so become the counters in the mediation. The nature of a particular system of governance will thus depend on endorsed and relevant propositions about the human personality.

If all men were angels, governance would hardly be required. Indeed, the "amount" of governance necessary to a society measures that society's conception of the void between men and angels. If man is, in part, of nature "red in tooth and claw," then governance alone endows the community with concern for the commonweal; left to his own devices—intelligence and ferocity—and moved only by his own appetites, man would speedily reduce his kind to a society of pharoah and fellaheen. If, on the other hand, man is an unperfected angel, then governance repairs in part that imperfection and permits human society to flourish as a crude surrogate of the society of angels. In the former case, governance provides the wisdom to perceive the longer and greater good and exercises the discipline to keep that perception from being overwhelmed in the competition among rampant egos. Governance viewed from this perspective seems, according to Royce Hanson's analysis of the destructiveness of private development institutions in metropolitan regions, to have broken down for the metropolitan community; governmental reform will thus mean restoring the capacity for collective foresight and individual restraint. In the latter view, governance generates information about the preference ordering of collective processes for achieving consensus about collective ends and means. Thus, doubting the ultimate perfectibility of the institutions of metropolitan governance, Julius Margolis sees reform as diagnosis and prescription for resolving specific and analytically distinctive problem syndromes. "Good governance," then, can be either that which exemplifies social vision and mobilizes social energies to follow the wise path, or that which is most sensitive in detecting and providing what its constituency wants in the way of goods that can only be enjoyed jointly, according to the accepted idea of man in society.

Ideal forms of government thus grow out of compelling conceptions of the ethical nature of man, and to the extent that the reform of governance attempts to merge the real with the ideal, its content and rhetoric will be shaped by the power of these conceptions over the behavior of constituencies. The making of the U.S. Constitution represented the triumph of one concept of man in society over another, the idea of the king as steward, in God's behalf, over the welfare of society giving way, after two centuries of struggle, to a concept of man as the measure of all things and government as the instrument by which he pursued his own ends, if not, indeed, his own salvation. Thus, competitive ideals of governance may ultimately supplant older ideals and give rise to new governmental responsibilities and activities, and, ultimately, to new governmental forms.

What has this mini-essay on the nature of man and society to do with the reform of metropolitan governance? It is only to suggest that the conscious modification of the institutions and processes of government—or "reform"—can come about as a result of *two* kinds of change. How governmental institutions perform may change as a result of how those institutions relate to objective characteristics of the real world. As Margolis points out, for example, the rich may segregate themselves from the poor, thus reducing the resources by which the public needs of the poor are met and so creating a serious political demand for institutional changes to prevent such unilateral decisions; growth may create such intolerable burdens of congestion and may so erode the quality of public goods produced that a large share of the constituency will insist on some other allocation of the burden; or new financial resources may become available, thus enlarging the potential role of government. Melvin Levin's evaluation of metropolitan Boston's experience in confronting new responsibilities to produce a new bundle of public goods—an amenable environment—is an illuminating analysis of how the government of a great metropolitan region has responded to a new set of roles.

However, another source of reform involves change in dominant constituency views about the appropriate role and responsibility of government in the lives of men. Hence, reform involves not only assessment of the performance of a system of government and a search for means of improving it, but also redefinition of performance and a reordering of the criteria against which it is to be measured. In these terms, Letwin sees the Royal Commission on Local Government challenged to find some arrangement of governmental institutions that can strike a new balance between the principles of efficiency and democratic participation; so Hanson sees governmental reform as "an exercise in the goring of oxen" which greatly need goring if larger public purposes are to be served. Ideally, the question reformers with the long view should be trying to answer is not only how objective characteristics of governmental activities are likely to change in a definable future, but also how people's expectations of government are likely to evolve. And, indeed, they do evolve. The New Federalism, with its fiscal *impedimenta* of general and special revenue sharing, suggests the emergence of a new view of the relationship of a citizen to his society. While it seems to return the initiatives in the political system to the citizen in his relationship to his state and local governments, other considerations vitiate this view. First, local discretion in the disposition of local fiscal resources appears to be enhanced, but at the expense of the coherence of *legislatively declared* national policies to alleviate poverty, to disseminate civil rights, and to distribute more equally the benefits of economic growth and change. Second, the renewed emphasis on extant civil jurisdictions of the municipal corporation and the state tends to ignore the growing host of issues we have come to call collectively "the metropolitan problem." Finally, the emerging field organization of the federal government suggests not a weakening, but a strengthening, of the capacity of the *executive branch* of the national government to leave its imprint on local policy. The new view of the relationship of the local citizen to his political environment, hence, sees him as ultimately parochial, indulged (or deprived) by the faceless formulas of revenue sharing administered by a conservative, if not inert, governmental system. Reform and change will become increasingly the responsibility of the courts, until, and unless, they too succumb to the proposition that the American system of governance has no responsibility for foresight, equity, or the claims of the disadvantaged. Although it is not clear how widely such a view is understood or held, recent trends seem to be in that direction.

In the longer run, those efforts toward governmental reform are most likely to succeed which take into account the demands emerging in the political constituency; it is in this context that the following essays should be read. They do offer suggestions for change, but in each case we must deal with the underlying question of consistency, not only with the analytics of the metropolitan problem but also with the expectations and aspirations of men in their relations with each other in the United States in this last third of the twentieth century.

If the reform of metropolitan governance is fundamentally a process of discovering a more fruitful relationship between a political ideal and the realities of integrating what public goods constituencies desire with production possibilities, such a relationship will be expressed in the way in which governmental institutions are organized to do what constituencies want them to do. The politically sophisticated often maintain that reshuffling organizational tables creates the illusion without the fact of reform; certainly much of our experience at governmental reform bears this out. Will the housing code really be better enforced by the city inspection department than by the public health office? Does it make any difference to anyone whether a regional planning agency is responsible to or associated with the council of governments instead of constituent local planning organizations? Are there welfare gains to the larger community to be realized by transferring the

National Forest Service from the Department of Agriculture to the Department of the Interior?

Obviously the answers to such questions range from "yes" through "maybe" to "no." In some cases an affirmative answer must be based on the prospects for better management, in others on more effective implementation of a particular substantive policy, in still others on the capacity of a change to embody a new definition of government. Only the last case emerges from, and is affirmed by, a true sentiment for reform, in which the aspirations of citizens who perceive new, more rewarding relationships with their collective institutions are creatively expressed in the realignment of the institutions of governance, and a reallocation of collective responsibility and resources. Thus, although reorganization does not necessarily mean reform, there will be no reform without a creative restructuring of the institutions of governance.

This volume addresses the organizational aspect of the reform of the institutions of governance of metropolitan regions. An intricate context of evaluation must be brought into play when, in the name of reform, governmental institutions are wrought into some new configuration of the political ideal and the social reality. Much of its value will be missed by the reader with a more mechanistic or manipulative view of organization, for he will fail to grasp the subtle nature of the medium we must use to approximate, within the limitations of the human condition, the society of angels.

My purpose here is to point to an analytical core for the papers in this volume: organization is an expression, not the reality of reform, and reform itself an assertion of a new relationship between the ideal and the real of political life. In this light, Royce Hanson examines where pluralism and political complexity have brought us in terms of the ability of metropolitan governmental institutions to shape the region's development in a manner consistent with the welfare of its constituencies. He contends that political fragmentation has served well the ends of a powerful and complex metropolitan-wide development industry, reducing those victimized to inarticulate protest. It is no wonder, as Hanson sees it, that stronger systems of regional governance have been fiercely resisted by the private interests who profit from land development. Such an arrangement of private weal and public woe will not be transcended easily, and probably not without the aid of allies of reform from without: the state, assisted by the federal government, can create planning institutions, implement measures, and encourage reforms necessary to make metropolitan development serve the interests of a metropolitan citizenry.

Julius Margolis inferentially doubts the existence of a metropolitan constituency in any sense other than the polemical, and he has noticeably strong reservations about the ability of metropolitan general government to detect, interpret, and respond to the preferences of the complex set of constituencies that compose the metropolitan political economy. Certainly there is no consensus about the optimal organization of government in a general sense, and without that consensus, there can be no general rules for reorganizing government. However, the metropolitan political economy is characterized by definable failures of the market system, for which there are specific prescriptions, many of which have strong organizational implications and any of which can reduce the urgency of general structural reform. Margolis, as pragmatic reformer, puts together an articulated set of prescriptions for improving the well-being of the metropolitan political economy, happy to push further down the road, if not out of sight, the imperative for general reform.

Melvin Levin examines specifically a metropolitan governmental system confronted with a substantial change in the demand for public goods in the form of an improved physical environment. Evaluating the response of government in a definitive fashion at any particular moment is difficult, but obviously it is more difficult at an early stage, when significant consequences may not yet have become visible. Recognizing this limitation, Levin makes the best of a difficult job by examining several environmental roles in the context of the responsibilities of a number of governmental agencies whose activities touch on the environment. On the "supply" side, he defines the metropolitan problem as one of the distribution and behavior of populations and economic activities within the metropolitan space, the regulation of which is the prime purpose of government responsible for producing these environmental goods. Lack of coherence among governmental institutions, together with the complexities of environmental and ecological processes, has resulted in low marks for the performance of Boston's governmental system. In the end, Levin argues, not new metropolitan governmental institutions, but the state government, offers the solution—the integration of metropolitan planning processes and consistency among quasi-independent agencies in the production of these new public goods.

William Letwin examines carefully the logic of the reform process in Britain as expressed in the extended study of the Royal Commission on Local Government (the Redcliffe-Maud Commission), published in 1969. The dependence of effective expression of the preference of political constituencies on the "smallness" of the political jurisdiction inherently conflicts with the efficiency advantages of "largeness" in the production of public goods, and, he contends, reforming governmental structures involves discovering the best way in which to mediate these conflicting imperatives. In the end, the Commission recommended a multitiered system with small participatory (but hardly sovereign) local elements, a larger set of fundamentally governmental jurisdictions, and, transcending these, five great regional institutions whose fundamental role was one of planning and foresight. A unitary system can, of course, thus sweepingly reform the entire subnational system of government and expose its logic for public scrutiny, but this lesson loses something in the translation from the British experience to that of a strongly federal system such as ours. It does, however, make a point often overlooked in our own experience: there is a strong *national* interest in the way in which metropolitan regions are politically served that needs to be introduced in the discourse about governmental reform, because the impact of reform on the interface between the constituency of one metropolitan region and that constituency composed of the rest of the nation can encourage or retard fruitful relationships between the two.

It is not, then, so long a step as it first appears from the mundane concerns of these papers with a vividly present political reality, known popularly as "the metropolitan problem," to loftier speculation about the ideal in the affairs of men. It is the latter that bestows spiritual vitality, defines practicality, and puts forward images of what is possible, all in a way that distinguishes the step upward—reform—from sideways movements and downward slips. For the true problem of reform is recognizing it.

1 Land Development and Metropolitan Reform

ROYCE HANSON*

Development of land is the most obvious manifestation of the forces of metropolitanism. The impact of land development on important metropolitan functions has for many years been a major stimulus of initiatives for metropolitan reforms. The unsightliness and inefficiency of suburban sprawl, the poorly distributed burdens of rapid population growth concentrated in jurisdictions with inadequate fiscal resources, and the growing disparities in financial capacities among neighboring jurisdictions resulting from concentration of property values have provided major justifications for proposals for metropolitan governmental reforms. But metropolitan reforms aimed at supplying the infrastructure to serve development—port authorities, water and sewer districts, parks authorities and transportation agencies—not only serve development; they generate it as well, and not always deliberately. More recently the metropolitan reform movement has sought to influence the pattern and type of land development through planning or coordinative organizations. But strong political resistance to introducing general-purpose metropolitan government has reduced the movement to relying upon such politically inoffensive organizations as metropolitan planning commissions and voluntary councils of local governments to deal with the powerful political and economic forces at work in metropolitan land development.

Other kinds of policy reforms have also sought to influence patterns of land development in metropolitan areas: redistributing interjurisdictional revenue can have a major impact, making possible more rational location

*Chairman, Montgomery County Planning Board; Maryland–National Capital Park and Planning Commission.

among metropolitan jurisdictions of major economic activities, such as employment centers. State development corporations with land assembly and development powers can strongly influence not only where development occurs, but what type it will be and what mixture of uses and income it will exhibit. Federal grant-in-aid programs have also increasingly demanded a minimum level of metropolitan planning to coordinate federal programs aimed at alleviating urban problems.

The federal role is the best developed of these recent policy changes. The 1962 amendments to the Federal Aid Highway Act were designed to minimize the negative impact of the interstate highway program on the development of urban areas. The more recent requirement for a central, metropolitan-wide review of federally funded, local projects is designed to better coordinate a broad range of federal programs that affect development and to nudge individual metropolitan areas toward more unified urban development policies.[1] The national land-use policy legislation embodied in Senator Jackson's 1973 bill was also based on the assumption that metropolitan development patterns were using land inefficiently from a national and regional point of view.[2]

Land development, then, is a central issue in metropolitan reform. How the location, character, and rate of land conversion are managed will largely determine the distribution within the metropolitan area of the population by race and class, the capacity to provide for housing requirements, the generation of traffic, the production of sediment, storm water, sewerage, solid wastes, and the demand for water, community services, and public expenditures. As a major economic activity, at the very heart of local politics and government, land development is an institutional problem involving systematic patterns of behavior by people, firms, and public officials. Any metropolitan reform that would alter land uses or development by making it more rational in metropolitan terms must, therefore, change existing power relationships. Conversely, any reform that would conceivably alter the distribution of power in the metropolis threatens the operation of existing land development institutions.

The Private Institutions of Land Development

Land development is predominantly a function of the private sector working within a complex network of governmental regulations. The multiplicity of private land development interests has preserved the competitive character of the land "market," whose major actors are land wholesalers, developers and builders, financiers, land retailers, suppliers, fixers, and consumers. Each

---

[1] See Office of Management and Budget Circular No. A-95.
[2] S.268, U.S. Senate, 93rd Cong., 1st sess.

group's interest in the land development process differs somewhat from those of the others.

The wholesalers are owners of urban land that is ready for conversion to more intensive uses, such as farmers at the urban fringe, and the real estate "packagers" and speculators. Normally, the packager works on behalf of a specific client who is seeking a suitable site for development. He may, however, operate on his own in a "ripe" area, seeking to assemble land for prospective clients; in this guise he shares the characteristics of the speculator, who is more often engaged in long-term investment in land, whether or not he has a specific use or client in mind. He relies on his ability to sense and exploit changes in land values under the pressure of urbanization to yield greater net returns than he would receive from other investments.

A developer purchases land to improve it for sale or lease in a short period of time to other builders or consumers. He may actually build on high-yield land, where retention of an equity interest offers a better return on investment than sale: shopping centers, office buildings, industrial parks, and apartment buildings may thus be more attractive than low-density housing. Builders include a wide variety of businessmen—single-lot, custom-home builders, large subdivision, apartment or office builders, and general contractors. This interest group also includes the building trades unions.

Development financiers, who decide what gets built, as well as the financial terms under which building occurs, are often the most influential factors in the market. The financiers are highly specialized. Major investment organizations, such as insurance companies, union trust funds, and large commercial banks, invest heavily in land development. Mortgage bankers put together financial packages to support acquisition and development of land or construction of buildings. Savings and loan associations and smaller banks provide the major portion of consumer mortgage money.

Public financial institutions, such as the Federal Housing Administration (FHA), the Federal National Mortgage Agency (FNMA), the Federal Home Loan Bank Board (FHLBB), and the Community Development Corporation (CDC), are also critical parts of the financial organization of the land development market. FNMA regulates the amount of mortgage money available and therefore influences rates of interest charged to builders and purchasers, and especially the rate at which the market can "move." FHA and FHLBB influence development practices, site planning, pricing, and home mortgage practices. CDC, the newest federal agency affecting development, guarantees loans for new communities and is often the determining force in deciding whether development will occur in a cohesive, well-planned package or come in the usually incremental, disorganized pattern.

The term *fixers* is not pejorative. Lawyers and lobbyists are essential to land development. Land-use law is a complex and arcane art. The lawyer and the title company are necessary to clear titles, to negotiate purchase contracts and financing, to establish and dissolve corporations, to obtain zoning and other needed development permits, and to sue and defend on behalf of their clients. In many respects, the land lawyer is the pivot around which the development process seems to revolve. He is involved in the economics of the development, its business organization, project planning and design, public regulation, and customer or public relations. As a result, the relatively small number of attorneys who specialize in land-use law wield great influence in the overall process of development, shaping public policy and their clients' political perceptions of the metropolis in which they work.

The lobbyist performs a less significant but still essential role. He looks after his particular segment of the "industry"—homebuilders, bankers, building trades. Although his focus tends to be narrower than the lawyer's, he uses some of the same skills to produce public policy beneficial to his special interests: amending building or plumbing codes to exclude competitive materials or groups, increasing the flow of money, reducing interest rates, changing zoning laws, or setting up internal standards of conduct for homebuilders or realtors to increase public confidence in the industry.

Naturally, the interests of all the forces in the market do not coincide. Fixers are often busy trying to outdo competitors or to find a basis for alliance and compromise on mutual interests. Land retailers (e.g., real estate brokers) maintain an interest in an active and expanding market, favorable rates of interest, and a political climate conducive to growth. The suppliers of building materials and furnishings find their major source of new income in market expansion, which means a constant demand for new appliances, furniture, telephones, paint, and electricity. The interests of all these groups converge when it comes to growth. They may compete for shares of the market, but in a rapidly expanding market all can be accommodated.

The other end of the market equation, the consumers of the final products of land development, have a more complex set of interests. Customers for new homes and business space generally benefit from high rates of growth. But settled business in older areas of the metropolis and people living in already developed communities may oppose new development. Small business, which often operates on a narrow margin, is frequently hostile not only to commercial area renewal programs, but even to street improvements, which create a short-term reduction in trade. Homeowners fight new development that will increase school enrollment, traffic congestion, and taxes, or reduce amenities. Since almost all development has some such effects, homeowner groups and occasionally renters frequently oppose any new development near enough to have political impact. Other consumers may ally with "developer" interests when they sense that development will improve the tax base. Consumer interests have an important influence on both the type and location of specific development activities.

The complexity of the major groups making up the land market assures that it is, to some degree, pluralistic rather than monopolistic, with many characteristics of a free market. But all of the actors are human, and the critical market processes involved make it clear that the land development market is a social, as well as an economic institution.

Before land can be developed, it must be assembled, an almost exclusively private activity. By the time a metropolis has become "self-conscious," most very large tracts of land have already been developed. In most parts of the country, small tracts have been developed without too much relationship to each other: the original landowner's subdivisions represented a sales plan, not a development plan, and as pointed out above, those who sell land often do not develop it or build on it. Now that the age of the metropolis has arrived, the land ownership patterns of the past are a heavy burden. The strong addiction to property rights in the United States gives the owner of each parcel the right to underuse, misuse, or overuse his land within very broad legal constraints. He is not compelled to sell it, so long as he can pay taxes on it; he can abandon it; he can will it to his cat; he can hold out as long as he wishes against any private effort to assemble, whether because he wants more money, or wants to stay where he is, or just dislikes the would-be purchaser.

This system of land tenure greatly increases the cost and time, and therefore the risk, involved in land assembly. Since assembly at the periphery of a metropolis is cheaper and easier than near the core, new housing development, shopping and office centers are built there, leaving behind older areas and generating new demands for public services and for supportive land development. Fragmented patterns of ownership make private assembly for major intown redevelopment or for satellite new towns difficult and costly. Indeed, the location of new towns in metropolitan regions is dictated entirely by the availability of a willing seller, not by the suitability of the location for that particular type of development. Development thus occurs in an almost random fashion.

Development financing is a psychoeconomic process: while investors compete, they also communicate and closely observe each other's mistakes, and not always with positive results. For example, some areas of a metropolis are not considered good investments, for reasons that appear to have little to do with economics: indeed, the commonly held belief that the older business districts and large parts of the inner city are unsound investments is frequently a self-fulfilling prophecy accelerating their decline.

The pluralism of the market also produces extravagant investment cycles in different land uses. A strong market for office construction is very likely to result in overproduction. High vacancy rates persuade investors to avoid loans on office construction and may result in excessive residential building. Lenders hesitate to accept new types of development until they are virtually riskless, like condominium apartments. The image of cold economic calculation is frequently obscured by the comic opera aspects of market decision making.

In large, growing metropolitan areas, it is difficult to lose money on land investments. But most money for land development and construction is available on a short-term basis. The terms of lenders and the thin margins of builders put the emphasis on projects with short build-out times, since the loan must be repaid by sales.

Marketing of land and buildings, particularly housing, has historically been a bastion of institutional racism and economic discrimination. Building and selling single-class communities was believed to reduce risk, open marketing to reduce property values. Reinforced for years by FHA policies, housing marketing contributed substantially to racial and economic segregation in metropolitan areas. Since the enactment of the 1968 Fair Housing Act, much of the overt discrimination has ended, but substantial vestiges remain. Many newspaper advertisements still carry such code words as "exclusive neighborhood," or even proscribed racial phrases, while affirmative advertising of equal opportunity remains limited.[3] Mortgage lending institutions still tend to regard minorities as higher risks than other home purchasers. There remain many examples of selective marketing—showing some neighborhoods predominantly to blacks or whites—and block-busting still occurs. Real estate sales forces are not well integrated. Black realtors are a rarity in the suburbs, and minority people in the home loan offices or on the boards of directors of lending institutions rarer still. Thus, the marketing system for homes heavily influences the distribution of population groups in the metropolis.

The arbiter of the private market (at least in theory) is the price system, but the structure of land prices is greatly influenced by public decisions regarding land use. Land prices depend on the availability of land for development—what is for sale and whether it is properly zoned or rezonable. Residential land is generally purchased on the basis of its "yield" in dwelling units. Thus, increasing density of "buildable" land is associated with rising market value. Nonresidential land in any given zoning classification may vary greatly in price, depending upon its intended use, the amount of similarly classified land that is as well located, the short-term state of the nonresidential land market, and the amount of marketable floor space that can be placed on the site.

Land must also be served by sewers and roads to be available for development. Sewers are particularly important for development of any density, and sewer availability can more than double raw land values under the same zoning classification. Civilization and prices follow the sewer.

[3]George W. Grier, *Bias in Newspaper Real Estate Advertising* (Washington, D.C.: Washington Center for Metropolitan Studies, 1970).

Location or accessibility is usually more important in the price of land intended for nonresidential development. "Hot" areas (where prices are extravagantly high) are often creatures of special circumstances, such as the concentration of particular uses, or access to a high-quality retail market. Communities with reputations for good public services are generally considered better "locations" and command higher land values than neighboring areas with perhaps better geographic characteristics.

Land prices are also influenced by "plottage," the ability to assemble a suitable amount of land, which affects yield and construction costs: an acre of high-density residential land is worth more *per square foot* than a few thousand square feet because the floor space that can be constructed on the larger site will provide a better return on the land. Carrying costs are also important factors: the price of land is its sales price plus its cumulative discounted value for taxes and interest. Land for which no market can be seen for several years will very likely remain undeveloped or be sold for a lower use at a lower price.

These factors and others, such as topographic conditions, which limit the yield in units or floor space, combine to set the price of any given parcel of land. Many of them involve nonmarket or noneconomic elements. Zoning, for instance, is a heavily politicized process in which conflicting interests often weigh at least as heavily as the appropriateness of a particular use on a given piece of land. Location may be as much influenced by psychology and politics as by raw economics. To the extent that price reflects all these forces, however, it is their one economic measure. The problem is that not too much confidence can be placed in the price system as a measure or arbiter of the market.

In summary, the market operates as a metropolitan institution, especially as regards major land uses like employment centers and office and shopping complexes. This metroeconomy is greatly influenced, however, by the policies of local jurisdictions and by the pluralism and specialization of the private individuals and firms participating in the development process. While extensive diversity of interests in market operations provides substantial competition, an oligarchic pattern of political influence within each governmental jurisdiction involved works in the opposite direction. Because those interested in land development frequently operate as the single most cohesive business group in local politics, local governments generally reflect the conventional wisdom of the market.

The "market" as an institution of the political economy has produced an incremental, outward movement of land development due to the difficulties of alternative assembly opportunities and to the inertia of expanding public facilities incrementally at all points rather than "favoring" one part of the market by selective expansion in certain locations. The market, as the governing force in land development, is characterized by short-term optimization largely constrained by financing and marketing considerations. Land development decisions (exhibiting significant externalities) are optimized by firms and individuals rather than by regional or local governments. Consequently, there are no institutional structures where the externalities of development decisions can be taken into consideration.

In the past several years, however, the development industry has moved steadily toward concentration. Large national corporations and conglomerates—ITT, Gulf, Humble, Westinghouse, American Cyanimide, and Boise-Cascade, to name only a few—have been absorbing small, individual firms or establishing their own building or land development subsidiaries. The entry of major corporations into an industry formerly dominated by essentially independent local firms promises to reduce substantially the pluralism and specialization of the private side of the market and to increase the oligarchic politics of the market. Although concentration could produce a more favorable political climate for regionalism, since the new firms operate at least on that scale, it could also reduce the marginal influence of public policies on market operations, since large conglomerates can internalize almost all critical aspects of the market, from land assembly through financing, development, and sales. A major national corporation, for instance, may already own land, originally purchased for mining or some now obsolete activity. An insurance company or other investor with a heavy cash flow and long-term investment capability may be among its subsidiaries, providing a steady flow of capital at favorable interest rates. The development company in such a case not only keeps the parent corporation's land and money busy, but opens up markets for its principal business, such as lumber, electricity, or petroleum products.

The Public Institutions of Land Development

The market cannot be understood fully except as it interacts with the public institutions of the metropolis—the governmental system, the legal and regulatory processes, the fisc, public improvements, and the planning process—through which officials participate in land development.

The Basic Governmental System

With few exceptions, American metropolitan areas are not governed by coherent public institutions that allocate public resources or provide services at the metropolitan level. Instead, there are many layers of special or general government, each dealing with some fragment of the political system. This system is important to land development, because it structures power and access to power.

The general local governments in metropolitan areas, developed under general state constitutional or statutory law, were designed for the most part for the pre-automobile age. Since these local governments tend to embody the fundamentalism of the "miniature republic," they have not been careful to provide consistent systems of law governing land use. Even large jurisdictions, such as the urban counties in the Middle Atlantic states, may maintain very different zoning classifications, tax policies, building codes, and policies toward land development. These differences may reflect interjurisdictional competition or simply local prejudices concerning the type of development desired. In any event, the metropolis normally is not constitutionally organized to permit the establishment of a metropolitan constituency or to require the reconciliation of contradictory local policies.

From the point of view of land development, the functional or special-purpose agencies and intergovernmental bureaucracies may be as important as the general-purpose, local jurisdictions. Major public works are often the responsibility of independent, special-purpose authorities. Each such agency deals ordinarily with only a very small part of a continuous and extensive process, and the lack of coordination among agencies frequently adds to the time-cost of development. The mutual independence of such agencies also occasionally produces contradictory policies among special districts as well as general-purpose jurisdictions.

The political fragmentation of a metropolitan area places the management of public activities in the hands of bureaucracies over which no political body can exercise the surveillance necessary for political accountability. The regional office of FHA answers to the national office, the district office of the state roads commission to legislators and its parent state agency. The Corps of Engineers district is best reached through Congress. The water and sewer authority, often self-financed from special assessments and user charges, may respond to no one in particular. And where federal or state assistance is involved in an agency's operations, those who administer the grant regulations may oversee the activities of the agency. For the most part, these special agencies have direct contact with client groups, not with the public at large. It is not strange, then, that those interested in development take special care to cultivate the bureaucrats in the agencies, who naturally tend to care for "their" public, which in turn protects the agency from criticism or curtailment of operations.

The dispersion of governmental functions in the metropolis calls for processes to arbitrate among and coordinate the units of government. In the 1960s, the principal means of coordination was the council of governments (COG), a voluntary association of elected and appointed officials. As voluntary associations, COGs exercise no actual governing power. They have limited ability to persuade their component jurisdictions, through their clearinghouse functions under OMB Circular A-95 and through their advisory metropolitan planning functions. Essentially, they act as communications mechanisms for local officials and channels for interbureaucratic negotiations. They generally cannot arbitrate or reconcile differences among different member jurisdictions, although they may provide a neutral ground for combat and compromise.

In many cases, the special-purpose authorities are supposed to be both the advocates of public interest and the judges of disputes between private and public interests in land development. In practice, such a conscious separation of functions rarely exists. Within a single jurisdiction, the local council may be able to decide a dispute between a public agency and a private interest, but there are no arbiters available at the regional level. In any event, the state courts are still heavily involved in settling land development issues on the bases of statutory and case law that may be some years out of phase with local or regional development objectives.

In the modern metropolis, then, the critical processes of government—aggregation and representation of interests, resolution of conflict, regulation, taxation, capital improvements and public services, and planning—each take place within a specialized institutional setting. The governmental system is characterized both by an areal and functional diffusion of power as it relates to development, and by competition among local jurisdictions rather than cooperation on matters of regional significance.

The coordinating processes that do exist occupy only part of the time of local elected officials, most of whom serve only on a part-time basis. It is thus extremely difficult for a metropolitan public, aside from those with direct economic interests, to form and exercise effective political influence and power on metropolitan land development questions. This situation also greatly inhibits much initiative from the public sector in managing even those public processes which contribute to urbanization. For the most part, the public sector continues to rely on regulation as its principal means of participating in land development.

The Legal-Regulatory System

The laws governing land development in the United States draw on concepts of private property and police power. The police power under the Constitution provides the legal basis for zoning and other regulatory processes affecting land development. Nevertheless, deeply ingrained presumptions in favor of private property severely circumscribe the operation of zoning laws, which cannot, for instance, deny an owner permission to develop anything on his land.

Because it is based in the police power, the power to zone resides in the state or its instrumentality, usually the municipal or county government. This legal convention is strongly reinforced by the political mystique of home rule,[4] so that the zoning power is jealously held and exercised by each local jurisdiction, whatever the metropolitan consequences. Indeed, "home rule" charters enjoyed by local governments in most metropolitan areas permit, even encourage, land-use regulations that exclude low-income families from the local jurisdiction: large lot requirements, controls on density, and prohibition of modular and mobile homes by ordinance promote high unit costs and the homogeneous developments of small jurisdictions. The residual community of the poor is thus segregated and confined to central-city ghettos. Local restrictions on certain "undesirable" industrial and commercial land uses also skew metropolitan land-use patterns. Jurisdictions compete for uses yielding high tax revenues by passing favorable land-use regulations and offering financial inducements, and thus overzone for some categories of land use, such as retail, commercial, office, or industrial. Such actions contribute to the underutilization of such sites elsewhere in the metropolitan region and to the congestion of local traffic arteries.

As suggested above, zoning is an intensely political process, both within and among jurisdictions.[5] The final zoning authority is usually the local elected body, and, not surprisingly, those with interests in rezoning to higher intensities of use are often prime contributors to political campaigns. Local councils are cautious, however, about ruffling neighborhood groups adamantly opposed to rezoning applications. The fact that there is a body of zoning law should not obscure the even more relevant fact that zoning also constitutes a special arena of local politics, in which the law is but one implement of battle. Most states consider zoning to be an exercise of the legislative power, so the courts are reluctant to substitute their judgment for that of the local political body in particular zoning cases as long as reasonable standards have been followed and the "proper" outcome is fairly debatable. As zoning law has matured, these standards have become more obscure, and the case law on zoning has become a combination of substantive and procedural due process, with courts selecting between substantive and procedural standards in deciding cases.

Zoning is generally not related to local plans and planning processes. Although theoretically zoning is a tool for implementing plans, more commonly

---

[4]*Home rule* refers to the granting by states to designated, or specific classes of, municipal corporations the right to exercise certain legal powers without the need for specific authority from the legislature. Particularly, home rule allows a municipality to adopt its own charter rather than depend upon the legislature for its organic law, although some states specify the basic forms of government that charters may provide. The prevalence of home rule charters varies widely among the states.

[5]See Richard Babcock, *The Zoning Game* (Madison: University of Wisconsin Press, 1966).

the planning process becomes a means of rationalizing existing or prospective zoning. Indeed, most jurisdictions draft zoning ordinances years before they are politically able or technically competent to adopt master plans. Furthermore, a master plan is at best only advisory to the legislative body, even when an extraordinary majority is required to enact zoning in conflict with it. And, whatever status a local master plan may have in zoning matters, any regional plan has even less.

As a device for controlling development, zoning is further limited by the usual judicial requirement that all land in the same classification have applied to it exactly the same standards for use (Euclidean zoning).[6] While this requirement spells out for the landowner the setbacks, height limits, and other standards he must follow, it has encouraged bland and homogeneous land developments, poor design, and abuse of the land, all to maximize yields under the fixed standards.

"Floating" zones have been devised to meet some of the shortcomings of Euclidean zoning. These permit variable standards of development through review of the site plan by a planning or zoning authority, often in return for more intensive uses. Whereas Euclidean zones may be assigned by a legislative body through the comprehensive rezoning of a large area or applied for by individual applicants, floating zones normally may be applied for only by the landowner. The courts have allowed this arrangement by ruling that such zones are in the nature of special exceptions. Another device for increasing use is conditional or contract zoning, whereby zoning is granted an applicant contingent upon his agreement to build to a specific plan or within a specific time. Optional methods of development are also in increasing use, allowing "clustering" of development at increased yield in return for site plan review.[7]

All of these approaches seek to mitigate the crudity of zoning as a device for guiding development or controlling its quality. Zoning is essentially passive—a restriction on the uses to which private sites may be put, always allowing for some reasonable use. Zoning cannot dictate what will be built, or when, or where; that initiative depends almost entirely on the private actors in the market. Because of the incremental movement of the development market discussed above, most development activity usually occurs in outlying jurisdictions. It is these localities which, almost invariably, have the weakest

[6] See *Village of Euclid* v. *Ambler Realty Co.,* 272 U.S. 365, 71 L. Ed. 303 (1926).

[7] Such an approach was proposed for central business districts in Montgomery County, Maryland. See *Planning, Zoning and Development of Central Business Districts and Transit Station Areas* (Silver Spring, Md.: Maryland–National Capital Park and Planning Commission, 1973), the final report of the Citizens Committee to Study Zoning in Central Business Districts and Transit Station Areas.

systems of land-use regulation. Indeed, speculators, packagers, or developers may dominate in their drafting.

There are other serious problems in the legal-regulatory system. Because almost all zoning or subdivision cases are initiated by landowners, and because each application is decided on its individual merits, local authorities rarely review an entire area at one time in terms of such overall effects of development activity as environmental impacts, drain on regional energy resources, traffic congestion, housing supply for lower-income groups, school population, local service requirements, or impact on the revenue/expenditure ratios for the jurisdiction and its neighbors. Although subdivision regulations can be used to forestall development until specified facilities are in place, it is unlikely that development could (or should) be restrained totally, even if it would produce a substantial revenue deficit or other external costs.

In general, the institutions of land-use regulation are designed to reduce risk for the owner-developer. The public interest is supposedly protected by limitations on types of use permitted in each use class and by procedural due process. Nevertheless, serious questions persist regarding the balance of equities between private and public interests. The economics of legal and professional representation, for instance, give the land developers a powerful advantage. Such costs are absorbed into the overall development costs of the project, eventually to be paid by the purchaser of the developed land, a fact which gives the private applicant an advantage over both governmental agencies and citizen groups. Few local planning and zoning agencies have full-time legal counsel, and some lack professional planning assistance as well; citizen groups opposing a development often depend upon volunteer or inexperienced counsel. The record costs alone may render a full pursuit of appeals impossible.

The bureaucracies responsible for regulating land use and development often operate independently of the planning system, and of each other. The sporadic civic passion to "take politics out" of land-use regulation results in very specialized regulatory systems, remote from general public scrutiny, often responsive to clientele groups, and insensitive to community aspirations. Note especially the record of agencies responsible for code enforcement, such as inspection and licensing agencies. Seriously understaffed, they are also legally and politically underequipped to deal with overcrowding, deterioration of structures, and other code violations. Indeed, the recent experience of most major cities offers eloquent testimony to the gross incapacity of the institutional machinery to enforce land-use and building regulations. The regulatory institutions of land development normally do not contribute to the realization of the general development objectives of a local or regional policy, unless that policy is simply to make land available as the market requires it. They afford virtually no consideration of the public or external costs of development. Because each local jurisdiction regulates to optimize its own internal development objectives, serious consideration of the consequences of local land regulations on regional goals or welfare is precluded, and, given interlocal competition for taxable values, far more land is zoned for profitable uses than is required for regional growth and change.

Reliance on regulation as its principal means of participation in the development process places the public sector in a position of reaction rather than initiative. This regulatory psychology pervades public agencies responsible for administration of land-use ordinances and leads them to play either "neutral" or adversary roles in their dealings with the private groups involved in development, and, indeed, the neutral posture often disguises uncritical approval of most development proposals.

Furthermore, this regulatory psychology generates a climate of corruption, compounded of weak technical competence of regulatory staffs, ambiguous law and public interest standards, and the high time-costs of a drawn-out regulatory process in the face of competitive pressures in the marketplace. Suspicion of collusion between developers and public officials is pervasive, and the evidence of bribery and conflicts of interest is more than adequate to feed the suspicion. Over time, this suspicion hardens the regulatory system, makes it more rulebound and less flexible, and helps produce more uniform, expensive, but marginally more "virtuous" development. It tends, however, to withdraw the public sector further from an initiatory role in development, and to foster localism in regulatory processes.

The Metropolitan Fisc

Fiscal systems exist principally to raise and expend revenues, but they have a profound effect on metropolitan development. Metropolitan areas do not usually have areawide fiscal systems, but instead abide with an array of state, local, and federal fiscal arrangements. For land development, the most important characteristic of these arrangements is the use of the real property tax as the major source of local revenue. It is often the primary source of revenue for public education and usually the sole support for local debt service.

This reliance on real property taxation makes the amount, type, and pace of development critical to the fiscal and political stability of a jurisdiction. The need for revenues leads to attempts to attract land uses yielding high taxes; conversely, the need to keep the rate of tax increases low may lead to restricting development to those uses requiring few public services. The cost/ revenue squeeze generates interjurisdictional competition over the location of new enterprise—promotional activities, growth restriction, or zoning to encourage tax-desirable uses and discourage cost-undesirable uses. All jurisdictions will fight for regional shopping and white-collar employment centers or luxury apartments and subdivisions and try to avoid blue-collar industries and housing for large or poor families.

The actual effect of tax-oriented development policies, however, may differ from expectations, due to the vicissitudes of the entire fiscal system or of the use attracted. State aid formulas for education, based on need and local tax effort, as in New York, may produce a larger revenue credit in some townships for housing low- and moderate-income families than can be produced by a regional shopping center. The fiscal premise of urban renewal also seems unsound, in the absence of policies that assure both faster redevelopment and metropolitan relocation of families displaced from slums, to avoid simply shifting the "need" for services to another sector of the same jurisdiction. The folklore of tax economics and development, however, is the usual test, not sophisticated fiscal analysis.

The hard fact of urban fiscal life seems to be that taxes go up in any event. The assessable base may or may not increase with effective political demands that produce higher service costs, but heavy reliance upon the real property tax to finance local services does result in severe fiscal dislocations in metropolitan areas. Normally, the absence of a regional fiscal system makes rationalizing development difficult.

The Twin Cities Metropolitan Council has recently made some modest progress by obtaining legislation to regionalize a portion of the revenue from commercial and industrial property to permit equalization among jurisdictions and to help finance regional improvements. Recent court decisions regarding disparities in the tax base as it affects equality of opportunity in education may also counteract the historic pressure for tax competition through local development policies. It is more likely, however, that political pressure will hasten full state assumption of school financing and other services that fall before the challenge of an equal protection clause; basic local development policies will remain more secure than ever, and some jurisdictions will just have relatively more tax money to spend on marginal activities or to supplement the state minima for education.

Typically, property assessments rest more heavily on the value of improvements on a parcel of land than on the potential value of the land in its highest and best use. Thus, the assessment system discourages redevelopment, or even the upkeep of property, because a deteriorated structure greatly reduces the assessment of a parcel, even if it is in a prime location. Land-value taxation, or differential assessment of land and improvements, though theoretically desirable, will be difficult to attain in most metropolitan areas. The interest, both private and public, in the existing system is too great.

Assessments are closely related to zoning and therefore to speculation in assembly of large tracts and conversion of land. Since property is normally reassessed when sold, and since sale is often contingent upon, or closely followed by, rezoning to the desired use (the property being assessed at market value for that use), it is important to develop the land as soon as possible to minimize high carrying costs in taxes. This relationship between assessment and zoning can inhibit large-scale development, such as new towns, where a majority of the holding will not be developed for ten to fifteen years. Some jurisdictions avoid this problem by rezoning only when the developer is ready to build, adding to the uncertainty of both the development program and the ultimate use of the land. The possibility of developing a site in its existing zoning, rather than as proposed, has been a continuing problem for the development of Reston, Virginia.

Maryland, California, New Jersey, and some other states have experimented with various forms of preferential assessment, which tend to work well for development of a new town, such as Columbia, Maryland. There, zoning could be granted for the entire development, with the undeveloped portions remaining in farm use and assessed at agricultural land prices. In other than new town areas, however, speculators were enabled to acquire and withhold land from development at minimal tax cost while its market value escalated, either through rezoning or simple market pressure. In the meantime, the local jurisdictions lost immense revenues from grossly underassessed land.

Initially justified as a means of retaining open space through agricultural use, preferential assessment has instead promoted land speculation without producing any permanent open space. When accompanied by contracts to withhold from nonagricultural uses, or by escrow provisions whereby the jurisdiction obtains the difference in use and market value assessments at the time of sale, preferential assessments may be defensible. But studies of the system in Maryland indicate that its cost to urban counties is immense: one study demonstrated that the revenue forgone in one county could have permitted fee simple acquisition of 1 percent of all the land area of the county annually, with a much higher degree of control over development patterns.[8] The ideal would be a system of staged reassessment for new towns or other large-scale developments judged to be in the public interest, which would contribute to a style of development consistent with public policy.

In most jurisdictions, general assessment methods remain archaic. Reassessments occur periodically, and in rapidly developing areas substantial increases in assessments are recorded every few years, frequently with severe political effects, for a reassessment is a constructive tax increase. Tax increases of this magnitude generate political pressures that affect development: some will urge curtailment of any growth, others will seek only certain kinds of growth,

---

[8] Peter W. House, *Differential Assessment of Farmland near Cities: The Experience in Maryland through 1965* (Washington, D.C.: U.S. Department of Agriculture Economic Research Service, 1967).

and still others will seek to cut expenditures. The most likely target of efforts to cut expenditures is the capital budget, particularly new bond issues to support improvements serving new development.

Special assessments for capital improvements, such as sewers or roads, are designed to cover the costs of extending services to new development but in fact cover only the incremental public cost of the services. Thus special assessments may actually subsidize the developer. While location of such services as a sewer could be used to control development, the builder is often allowed to build his own system, deed it to the public, and recover its cost in the sales prices of his development. Or he may be allowed to "contribute" the marginal cost of service extension, and the public agency will build the facility. These arrangements not only facilitate the development in question, but often generate pressure for peripheral development or, if service has been extended some distance beyond the urbanized area, for additional access to the facility. In any event, special assessments or contributions in lieu thereof rarely meet the public cost of the service, and thus tend to add to both the tax and the development problems of a jurisdiction.

Other taxes have less direct effects on the pattern of land development. In spite of much discussion of the dangers of industrial flight from payroll, income, and sales taxes, little evidence supports the notion that taxes are a very important factor in location policy; services seem far more significant. Federal tax policy is more likely to influence the pace and type of development. Depreciation schedules influence the rate of building of apartments and offices. In some cases, federal tax loopholes can be made to yield returns from otherwise unprofitable ventures in land or buildings by proper management of tax losses. The federal tax structure, which provides little, if any, incentive for the upkeep or redevelopment of properties, may be regarded as a major contributor to the deterioration of the rental housing stock.

The cumulative effect of the fiscal arrangement of the metropolis contributes to the physical deterioration of older areas and the improper timing of much suburban land conversion, and sets up impediments to land assembly for such major projects as new towns or new intown development. The fiscal system encourages sprawl development and exclusionary zoning by local jurisdictions, and has the interesting effect of producing an uncoordinated alliance between some homeowners and some elements of the development industry, where both desire development that yields high tax revenues at low cost.

The Public Service System

The relation of public services to taxes and development was glimpsed above. The way in which public services are delivered in a metropolis is important to its development because in many cases public improvements are a necessary precedent if private development is to occur at all.

Second, the public is a major user of the land resources in a metropolitan area. Many public uses attract development, facilitate it, or even produce it. Airports and colleges, for example, are major generators of ancillary private development. Their location is of prime concern to local jurisdictions and private businesses. Mass transit, highways, parking facilities, and utilities all create land values and provide competitive advantages to certain areas. Renewal and housing programs provide business opportunity for some segments of the local economy and subsidize some jurisdictions.

Thus the location of public facilities and works affects land values, development schedules, and land-use patterns. Private development, in turn, generates demand for public improvements. The processes important for land development are public plans and programs for the extension of services, capital budgets, land acquisitions, and construction of facilities.

The public service system, if highly coordinated and coalesced into a metropolitan program and budget, could be a critical element in development policy. At present, the system is fragmented into specialized delivery systems operated by insular bureaucracies of state, local, and, very often, special-purpose governments. Each agency tends to have tunnel vision with respect to its own programs, and its political and professional officers rarely see its activities in terms of any broad development objectives. A transit agency may be cognizant of the pressures its station brings for adjacent development, and yet fail to coordinate fully its station designs with agencies concerned with related land uses.

Specific programs are normally understood in the same limited way—as expenditures or actions necessary to build a particular facility as an end in itself, not as infrastructure investment designed to achieve public policy toward development. This reflects a "service" rather than an "investor" orientation and role. An expenditure of several billion dollars in transportation has as its end result the operation of the transportation system. It is not seen as an investment in metropolitan development for which the public is entitled to any return other than the facility itself. Indeed, this myopia is so endemic in public services that vast sums of public capital are spent to assist private development with no quid pro quo expected in either social or financial benefits to the public.

Coordinating major public investments is hampered by the weakness of the capital budgeting process and the diversity of funding sources for such activities. With rare exceptions, general metropolitan capital improvements programming is nonexistent. At the state level, capital budgeting normally consists of a compilation of agency requests. In any event, there is no state capital investment program oriented toward metropolitan development. Even in the location of state facilities, little concern is shown for the developmental consequences of the action. Local capital improvements programming, still in its infancy, is often so politically fragmented that it is next to useless as an instrument of development policy.

In general, the public service system, instead of shaping or guiding metropolitan growth, serves private development. This system, over which the public exerts full control and which has profound impact on development, could be the cutting edge of metropolitan development policy through coordinated control of access, location, and services essential to land development; it has become instead the agency for private exploitation of the processes of urbanization.

The Institution of Planning

Planning—largely concerned with land use and closely related matters—has been regarded by some as the spearhead of metropolitan reform. Metropolitan planning is a federally sheltered workshop by virtue of the urban planning assistance program and the A-95 review process. In fact, the actual record of planning agencies exhibits little relevance to most day-to-day development decisions, but much to agency survival and to obtaining federal grants. Very few metropolises have political bodies capable of adopting regional plans and making them supersede local plans.

Local planning is still largely isolated from regional planning and from the executive and regulatory agencies of local government itself. As explained earlier, planning is usually regarded as advisory to local officials and regulators, and is not a self-implementing process.

The much-discussed gap between planning and action is the result of several institutional factors. The methodology of planning, particularly at the metropolitan level, is highly abstract or general in tone, deals with issues at a macro level, and has a long time-horizon, all of which seems irrelevant to the public official, whose horizon seldom stretches beyond his own term of office. Moreover, the goal-orientation and rational policy model of regional planning, which proceeds from analysis of individual elements through the use of inductive logic, is inconsistent with the pragmatic and intuitive approach of political decision makers, for whom the relevant fact is public opinion, not geology or computer models. As planning has become more highly professionalized, it has also become less participatory at the regional scale, looked on as the property of the practitioners of the black arts of computer simulation and model building.

Planning has little direct relationship to the development process. It is advisory to any capital improvements program and to the regulatory process, and the advice, especially in regional plans, is often too general to help with specific local or functional decisions. Operating agencies, threatened by comprehensive planning, jealously guard their right to plan independently for their own activities. Because functional agencies are often far better equipped in both money and manpower to do their own planning, the planning agency's role is frequently reduced to that of critic rather than initiator of policy. In the absence of a comprehensive metropolitan capital improvements program, "implementation" is also left to the very agencies most inhospitable to comprehensive planning.

At every level, and particularly at that of the metropolis, planning aims at "end-state" plans with no appropriate program of resource allocation or scheduling. Plans are, hence, simply documents of public persuasion, not *plans* in the sense used by business organizations or military groups. The acceptance of the separation of planning from implementation by planners seriously disables the planning process and limits its potential for development policy. If "development" were the end of the process, then planning would be the means of producing the public decisions needed to guide both public and private sectors to that state—not merely a delineation of what that state should be. As it now exists, metropolitan planning appears to be more an institutional capacity for fantasy than an integral part of development strategy. It is not surprising, then, that local officials are not heavy users of metropolitan planning products. This stems partly from the factors discussed above, but also from the nature of their legal duties, the demands on their time from an almost endless variety of competing official and private interests, and the schedule on which decisions about the budget, agency programs, or grant applications, for example, must be made. However, local officials have become more receptive to the idea and process of planning as it has gained public acceptance. Developers often remain hostile, however, if only because they perceive the planners' constituency as essentially anti-developer in interest.

The most important institutional characteristic of metropolitan planning remains the dominance of functional over comprehensive planning. Functional agencies, supported by clientele groups who come to their aid at budget time, generally have greater financial resources than the "comprehensive" agency. They can carry out their plans and can show results within the term of office of elected officials. And they use the compelling logic that they are service agencies just doing their job. Metropolitan planning, therefore, often consists of worrying about the unanticipated consequences for development of all the successful operations of functional agencies. The initiative is in the hands of the operating agencies and the developers, not of the planners.

Metropolitics and Development

The public and private institutions of metropolitan development, which structure its governance and channel the exercise of power, serve some interests better than others. The various participants in the development process, special economic interests and "citizen" groups alike, use the institutions available to them to achieve their objectives. Unable to get what they want through existing institutions, they may become interested in institutional changes, that is to say, in metropolitan reform. In short, the losers are most often the reformers, which partly explains why reform is so difficult.

Since private interests tend to align their own organizations with the official system, there is little point in metropolitan-wide organization where there is no metropolitan governmental body to influence. This arrangement substantially restricts the kinds of conflicts that can arise over development issues and structures the way in which conflicts can, or cannot, be resolved. Thus, a major developer can operate with greater license in the one locality where his economic activity is a large segment of the local economy than he can in a consolidated system of metropolitan government, where his enterprise is less significant and his influence diluted in a broader representative system. Moreover, since private development, whatever its regional impact, has to occur in some jurisdiction, a developer will be far more concerned with local development policy than with any regional policies, unless they affect his ability to develop. So long as the local zoning authority, for example, does not require conformity to regional plans or policies, such plans will not become salient factors in private development decisions.

Some interests—lenders, utilities, general contractors, and commercial center or office building developers—do operate on a metropolitan scale and have to think in regional economic terms. With their major industrial and commercial clients, they constitute an important political base for those efforts at metropolitan reform which will help them achieve their developmental objectives: they have supported better regional transportation, water, and sewerage systems; they have supported urban renewal programs to protect or enhance older downtown investments; they have sponsored "power structure" interest groups, such as the Allegheny Conference of Pittsburgh, the Greater Baltimore Committee, or Detroit Renaissance, which provide leadership and money for major developmental projects in both the central city and its hinterlands. The objectives of such groups, however, rarely require sweeping metropolitan governmental reform; in fact, general reform might introduce a political system that would accept or promote other priorities than a new international airport, a convention center, or downtown renewal. Since political cohesion among such groups depends upon specific attainable projects with tangible economic consequences for the business interests of the groups' members, their developmental politics seeks to limit any organizational reform to the least change necessary to effect their narrow objectives.

Seattle offers a case in point. Some years ago its water crisis threatened to stop development. Studies resulted in the adoption of an open-ended Metropolitan Authority with power not only to produce and distribute water, but also to discharge almost any other function it chose to perform. It chose to remain a metropolitan water agency. Only recently was Metro's jurisdiction expanded to include mass transportation.

Similarly, when regional mass transit has been accepted as a necessity, all the lip service to "balanced transportation" has failed to produce a metropolitan transportation agency with comprehensive power over the entire transportation system. The existing dispersion of power in a metropolis among various private groups and public bodies requires consensus politics, and consensus is always jeopardized when the sphere of influence of an existing agency or group is openly threatened. Such threats can be minimized by narrowly defining problems and the actions appropriate to deal with them.

This dynamic, reinforcing specialization of the institutions dealing with regional land development also provides a built-in clientele for the political support of such a *limited* agency, further enhancing its influence in the region. In such a fashion, government institutionalizes existing power relationships and often embalms an agency with a particular development philosophy. This process results in superior political access for some interests and very poor access for others, by which means the agency becomes to some extent a public instrument of private interest—a highly salutory outcome for those who are well served. Politics, like business, seeks monopoly, not competition, and the happiest arrangement for any interest group–bureaucracy alliance is a political monopoly over its arena, with an independent board and revenue base.

Where a clear political monopoly is not possible, a lesser stratagem is the creation of a *technical* monopoly. A frequent ploy of bureaucracies or private trade associations, the technical monopoly involves a constructive domination of the information or accrediting power required to make policy or participate in an activity. The most effective technical monopolies historically have been those exercised by highway agencies, which control both the production of data and the analysis of that data, on which decisions must be based. Another example is the sewer authority, which is the sole possessor of information on treatment plant and sewer line capacities: its data are necessary to other development interests in making decisions, which gives it considerable power. On the private side, realtor associations establish the conditions of membership and business practice and provide almost the only information available on current markets through such member services as computerized multiple listings.

Political and technical monopolies protect the interests of their holders and enhance their power in local and metropolitan politics. They permit strategies of limited conflict and contribute further to specialization and internalization of development activity, so long as they do not come into conflict with another existing or emerging group. Within local or special jurisdictions, there is rarely a broader polity in which the outsiders can seek coalitions to compete with the monopolies, particularly when the monopolies are officially sanctioned. Competition among the many interests in regional land development thus consists of comparative shopping for arenas by the respective competitors. In land development politics this results in a fairly complex appeals system involving a plurality of organizations with different constituent bases. In a single jurisdiction, a zoning case may receive a recommendation from the planning board, be decided by a zoning examiner, reviewed by the elected council, and appealed to the circuit and appellate courts before it is resolved. Each body may represent different political interests. That the system is untidy, time-consuming, and erratic is not contested. The point is that the pluralism exists because it serves particular interests well: neighborhood defenders who feel impotent before the planning board or zoning examiner can often exert strong influence on an elected council; developers who lose elections can still gain a "fair" hearing before special commissions and courts more specialized in development affairs. This complex decision process tends to keep development policy on an ad hoc rather than a comprehensive footing. Each matter is decided "on its own merits" without much reference to others. Offensive as this is to those with orderly sensibilities, it is advantageous to those with short-term, individualistic interests. It effectively obscures any view of the cumulative effect of development until after it has occurred.

Metropolitan reforms that would rationalize this system to an abstract concept of the public good are politically undesirable to those groups who seek decisions rationalized only to their own interests. In this context, general planning and comprehensive development policy has a very low level of political salience. Its constituency is small and not very powerful. Moreover, there is no metropolity within which a political base strong enough to overcome existing alliances can be built.

Development politics, then, often reduces to local conflicts between "developer" interests drawn from market institutions and "citizens" (i.e., homeowners). Historically, local governments have been dominated by development-favoring interests in the early years of an area's growth. In many instances, the landowners converting their land to urban uses are the leaders of the local party or of the government itself. "Development" is viewed as a public as well as a private good, and the task of government is to facilitate it.

As development advances, however, conflict arises. New residents voice dissatisfaction with schools, traffic congestion, and the impact of more development on their tax rates and the amenities of their neighborhoods, a conflict that intensifies as land scarcity drives up prices and increases the pressure on public agencies to permit higher residual densities. The new citizens increasingly demand tighter regulation of development and begin to fight for control of local political and governmental processes. Such battles normally ensue for two or three decades, the life of extensive development activity. During most of that time the development interests probably hold the upper hand, since they have established the rules of the game through their early, virtually unchallenged control of local affairs, and have thus institutionalized many processes that must be altered if a change in development policy is to be effective. Development interests are politically cohesive and capable of concentrating their political activity. Where party systems are used for nomination of local officials, it is necessary only to control a majority of the nominees in the dominant party. Given the cost of campaigning and low primary election turnouts, money is an important element in the nomination process, and, as noted earlier, the principal sources of campaign contributions are often the development interests.

"Citizens" are hardly of one mind or interest. Specialization of political interest is pervasive. For most, controlling development is not nearly as important an issue as taxes, schools, or other public services. Since candidates can hardly run on development issues alone, alliances are readily at hand that give those in power a good chance of staying there. It is not hard to transform the public image of those who would reform the development process into one of profligate spenders with little practical appreciation of where the tax dollars come from.

To the extent that political or technical monopolies have been established, change is made more difficult. Where independent state agencies have been established, even charter reform, a frequent nostrum, may lead nowhere insofar as control over development is concerned, because local charters cannot supersede state law. Control of local councils can help, but if the sewer or planning agency is an instrumentality of the state, then full control over the local legislative delegation may also be necessary. Such control can be gained only after a long period of political activity, and normally results in small, incremental reforms because of the need to compromise on the composition of election tickets and, subsequently, on the substance of legislation in order to ensure its enactment.

One of the familiar objectives of reform—"stop unplanned development"—is not nearly so easy to achieve. Planning takes time, and meanwhile development goes on under the existing legal and political rules of the system. Furthermore, producing a master plan to guide development requires agreeing on development values, holding public hearings, and taking official action. Reform groups are more often agreed on the need for planning than on what a plan should contain. Once again, opposing alliances can effectively delay action, giving the advantage to those who have fared well under the existing regime. What normally emerges are mechanistic responses to organic problems. New procedures or organizations will be agreed to, but they are likely to involve no basic institutional change in the development process. George Washington Plunkitt's dictum that "reformers is morning glories" applies to the politics of development; few have the patience, the time, or the vital interest to endure the long struggle involved in institutional change. The culture of reform is also heavily oriented to organizational change, with no appeciation of the fact that organizations may be only a facade behind which basic institutions remain intact.

At the metropolitan as at the local level, the inertia of going concerns is the most important fact in the politics of development. The first law of any institution is survival; its ultimate skills are defensive. Officials of local jurisdictions stoutly resist metropolitan reforms that would abolish their offices or materially alter their ways of doing business; their opposition and that of their constituencies have frequently been decisive.

When change cannot be prevented, one defensive strategem is to foster harmless change. The council of governments (COG) is, in part, such a strategem for local governments, allowing the appearance of dealing with metropolitan problems through existing local governments. While COGs have been heralded as an evolutionary step toward fullblown metropolitan reform, most of the evidence leads to a different conclusion. Once established, COGs create bureaucracies that value survival above action and concentrate on "safe" issues on which intergovernmental consensus is likely. They resist substantial organizational change and emphasize the use of existing agencies and governments. Most conflicts involving COGs result from "turf" violations, moves by the COG bureaucracy to preempt activities currently performed by other agencies. Though COGs usually lose, they may come out of the fight with face-saving "cooperative" agreements with the agencies. The main strength of the COGs derives from the federal government, whose agencies regard them as political instruments to further their own program objectives.

The reality of institutional change is, therefore, that in both local and metropolitan settings it normally occurs through slight, incremental adjustments over very long periods of time. Normally these changes only parry or dull the thrust of reform and force the reformers to consent to "give it time to see if it will work." These modest changes, however, do set in motion new forces that often render the initial reform plan obsolete and require additional incremental adjustment.

Development institutions resist major changes because they reduce the predictability of the behavior of public and private actors in the development process and so raise the risk involved in development. Thus, the alteration in marketing practices required by fair housing legislation was strenuously opposed at first, and prevailed only after very substantial countervailing pressure was brought to bear at the national level. Few jurisdictions responded before the enactment of the federal law, and, since the act, change in marketing practices has been slow.

Because of the inertia of going concerns, most public policies affecting development have marginal impact and produce little significant institutional change. Indeed, most new policies tend to reinforce existing institutions and strengthen their ability to survive. Revenue sharing may well be the latest such reform. Conceived simply as a way of supplementing the revenue sources of local governments, it is not likely to provide leverage for rationalizing metropolitan or state revenue systems or for improving the capacity of local governments to deal effectively with the problems that generated the revenue crisis. Although such possibilities have been suggested, revenue sharing has degenerated into a means of subsidizing substandard governments in the performance of substandard services.

The experience of some metropolitan reforms, such as the creation of the Twin Cities Metropolitan Council and the New York Urban Development Corporation, suggests that major reforms may well have to be imposed by state legislatures, where a different set of political influences prevail. Only by leaving the metropolitan arena does it seem possible to put together the coalitions required for substantial institutional changes in metropolitan governance or development practices.

Whose Ox Is to Be Gored?

Writing constitutions and reorganizing governments is basically an exercise in the goring of oxen; in any metropolitan reform some interests will do better than others. The classic types of metropolitan reform—consolidation of local governments and federation—have more implications for the way in which development decisions are made than for the actual results on the ground. If representative systems are restructured, for instance, then regional capital improvements may better serve some groups now short-changed in the capital budgets of localities. If a metropolitan government has planning powers and can assert the integrity of its plan against the claims of localities, then a substantial change may occur in the location of regional facilities, employment centers, and large-scale developments. Metropolitan government produces a new political rationality in development policy simply because political interests are weighed differently in the metropolitan scale. This does not mean that development policy will be better, only that it will be different.

Elevating local development functions to the metropolitan level may well increase the influence of the large-scale development organizations already organized to do business on a regional scale and leave various citizen groups less able to veto specific projects. Minorities may find allies, or they may be even further isolated from power. What is clear is that general governmental reform offers little that is specific in development policy. Its rationale is based on other grounds: namely, that it would provide a more democratic and manageable basis for the formulation and execution of a wide range of public policies affecting the citizens of the metropolis.

Whatever may be done about general governmental organization, other metropolitan reforms are needed to change development patterns, reforms that deal directly with the institutions presently governing land development. While general reform may be a prerequisite for some specific reforms, it is not for all. Reform or modification of market institutions is more likely to result from specialized state or federal legislation than from self-improvement projects.

Although general metropolitan reform is not the "answer" to the substantive issues of land development, it may be one element in an overall political strategy for the reform of land development institutions. The pluralism of land development institutions in the political economy suggests that any substantial reform will require a long-term strategic effort designed to alter the way in which the major institutions function. In this context, the effort to establish metropolitan government is important as an alternative means of authoritatively arriving at objectives for metropolitan development policy.

Comprehensive or radical reform of development institutions, however logical, is not likely, given the facts of institutional life. A strategy of reform, therefore, must be devised and pursued which exerts leverage on those institutions with the greatest existing or potential impact on the character of regional development—land assembly, financing of development, capital improvements, the fiscal system, land-use regulation, and planning. The following comments are no more than a point of departure for development of such a strategy.

Public Land Assembly

Probably the quickest return for the energy required to secure reform is to be had by altering the operation of the land market through public involvement in land assembly for development and redevelopment. Use of public development corporations can enable states, regional agencies, and even local governments to acquire strategically located parcels of land and to develop them in conformity with metropolitan objectives. The present trend toward concentration of the building industry provides a better climate for such action than has existed in the last decade. The only private enterprises capable of substantial redevelopment projects or multiuse, large-scale new development, are, increasingly, national corporations. The issue is whether to enable the public to decide where development shall occur by acquiring the land, or to permit major land development organizations to make that decision for them.

Development corporations, by acquiring strategic parcels outright or by assembling through joint-ventured, quasi-public corporations, can encourage builder competition by land leasing and sales of developed land, and can also reduce front-end costs for builders. They can also produce "balanced" communities by transferring into the public sector the trade-offs in land values or financing necessary to produce variety in housing costs and types and better associations between housing and employment. Such corporations could also make possible the "pairing" of inner-city redevelopment with new community development at the urban fringe.[9] This type of joint development can provide a means of using the economic leverage of urban growth to address inner-city employment, housing, and service problems. There is every reason to believe that public development corporations could not only pay for themselves from their total operations, but could replenish their revolving funds and provide means of assuring public benefit from development while minimizing its indirect costs. Although this might be achieved in some local jurisdictions by state enabling legislation, it is far more likely to work under state or metropolitan auspices, which provide greater financial backing and a political base from which to enforce locally unpopular decisions—condemning land in order to assemble it for a new town that could house some poor people and minorities, for example. In the American political system, a public agency exercising such necessary powers must be accountable to elected officials; where there are no metropolitan elected officials, the state has the only ones available.

Federal assistance to state and regional development corporations—low-interest loans in lieu of tax-exempt bond issues, for example—could help bring such agencies into existence, but their powers must still be determined by state constitutions or law. Alternatively, the federal government might become directly involved in land banking for urban development, though this seems politically doubtful except in support of the states or in experimental projects. Land assembly practices, however, if left to operate as at present, will dictate a continuation of incremental development and the bias toward urban sprawl and segregation ,of land uses and income classes. Large-scale development would allow a different pattern to emerge and would provide the coherently planned diversity necessary to accommodate critical urban needs.

Changing Financial Institutions

Financing development is already heavily influenced by public actions; in many ways it is dependent upon federal credit and tax regulations. Development corporations would alter the front-end costs of regional development by transferring the interest on land acquisition to the public corporation. Extensions of the debt guarantees or implementing the long-term loans to pay interest on front-end costs authorized by Title VII of the 1970 Housing Act would make large-scale, planned development more attractive. If such assistance is supplemented by state financing corporations, contingent upon development being located in conformance with regional plans, it will be easier to encourage development to follow such plans.

---

[9] A strong case for this approach is made in Metropolitan Fund, *Regional New Town Design* (Detroit: Metropolitan Fund, 1971).

Lower rates of interest resulting from public debt guarantees or direct loans (or even public development financed by bond issues) can result in financial leverage to require production of more low- and moderate-priced housing in new communities, and affirmative action programs for equal housing opportunities.

The success of Title VII in these areas strongly suggests that state housing finance agencies or new federal programs could profoundly affect both housing costs and the demographic distribution of the population by extensive efforts to reduce interest rates in return for such public benefits. State action on both assembly of land and financing of development could greatly alter marketing practices.

More subtle leverage on private financial institutions is also possible. Though marginal to overall development patterns when taken alone, such actions as partially basing state fund deposits on the social responsibility of the banks involved could be useful in a total strategy. Willingness to invest in low-income neighborhood redevelopment, a more enlightened attitude toward loans to financially responsible minority group members and firms, or cooperation in quasi-public development enterprises is certainly as acceptable a basis for depositing state and local funds as campaign contributions by bank directors. Some states, such as Illinois, have already made some strides in this direction.

Taxation and Development

Tax policy alone is not likely to induce development to succumb to the regional public interest, but it can be a tool of considerable marginal utility if used in concert with other devices.[10] Tax abatement or preferential assessment, for instance, combined with low-density zoning, easements, or granting to the public the development rights in critical areas, can reconcile public and private interests in some areas. Such policies can relieve tax pressure to develop prematurely. But they must be taken in total context, however, lest they provide a haven for land speculation.

Preferential, or, more accurately, staged, reassessment of large land areas planned as new communities can alleviate much of the pressure to develop prematurely, without proper regard for public benefits. Such assessment policy is, of course, biased in favor of new town development, whereas present assessment policy is often biased against it. To be effective, however, this sort of tax policy must be accompanied by state and regional policies that establish the substantive public benefit criteria by which private development could qualify for such preferential tax treatment.

State laws modeled on the Minnesota experience, regionalizing some, or even all, property taxes, could reduce interlocal competition for employment

---

[10] Carol S. Meyers, *Taxation and Development* (Silver Spring, Md.: Maryland – National Capital Park and Planning Commission, 1968).

centers and allow them to be located to better regional advantage, provided some intelligent regional planning is first conducted. Alternatively, state tax reform reducing local reliance on the property tax and sharing other revenues, such as a state progressive income tax, based partly on compliance with regional development objectives, could also affect local decisions respecting regionally significant land uses.

State Land-Use Controls

Regulation of development offers a more complex challenge to metropolitan reformers than do other strategies. The current effort to impose national or state interests onto a traditionally local process would be more encouraging were there any evidence that state susceptibility to private development interests was at least exceeded by some legal and substantive understanding of the policy quagmire about to be entered. But despite such qualms, state involvement appears necessary to avert a series of land-use atrocities ranging from destruction of ecologically sensitive areas, in the guise of establishing "recreational communities," to rampant subdivision without regard for the public consequences.

The presence of the state government as a force in land-use regulation introduces a new element into development politics. First of all, it may establish another arena for conflicts by providing a legal means of determining that the use or misuse of some lands is of statewide importance. This opens the door on the "critical area" in land-use law, where state restrictions that preempt local zoning and subdivision regulations may be imposed. The state may also become a party of interest in administrative or judicial proceedings affecting critical areas.

State intervention in land use might also involve preempting local regulations in order to develop state-sponsored low- or moderate-priced housing, new communities, or major employment centers. Where regionally significant development is concerned, the predominance of state or metropolitan policy must be made clear. If it is possible to coordinate classification of land and its taxation, then it may be possible to accelerate or retard specific development projects. No state has yet adopted general legislation regarding new towns, for instance, yet most local zoning codes are inappropriate for such development, which almost always requires substantial state capital investments, especially for roads. In many cases, new towns will be built in localities with no development codes at all and no technical capacity to administer codes. States should probably adopt and administer standards for new town development, unless the locality has adequate capacity. Location of new towns, especially, should be a state or metropolitan function.

Reform in local land-use regulation is essential in order to coordinate development with the ability of government or the developer to supply public facilities. This involves "staging" comprehensive rezoning to coincide with capital budgets and adopting regulations that allow localities to prohibit construction of "premature" subdivisions.[11]

Achieving so aggressive a role for states or for substate regions will surely generate opposition from local officials and result in either weakening the legislation that would make state action effective or diluting the administration of such legislation to the point that full legal powers are not used. In point of fact, if state administrations now used fully and aggressively their existing powers with respect to capital facilities and regulation of utilities, water quality, and land conservation, the need for new legislation would be diminished.

Expanding the state role in land regulation will give further impetus to state and metropolitan planning to provide the rationale for action concerning particular parcels or areas. One important facet of the state government strategem is that it will provide a focus where new metropolitan interests in economic, environmental, and social issues can register their opinions more effectively than at the local governmental level.

Coordination of Public Investment

Central to the achievement of public-oriented metropolitan development objectives are (1) channeling market activity to preferred areas and (2) coordinating public powers so that development can occur at those places, in an orderly manner, and in proper sequence. One way of achieving this is through the use of development districts—temporary managerial and service agencies established in accordance with state law and operating in locations consistent with state, regional, or local development plans.[12]

A development district would be set up specifically to coordinate the governmental role in the development of a particular area and would be abolished when the development phase was completed. It could produce detailed plans for the development of the land in its jurisdiction and sell or lease publicly acquired property to private builders to produce the structures called for. State or local governments could transfer to it the developmental functions of the governments' operating agencies. Using these powers, the development district could produce, in proper sequence, the roads, sanitary systems, school buildings, parks, community facilities, and services required to assure that the development will function properly. The district could be financed through a revolving fund for land acquisitions, transfers of capital

[11] *Golden* v. *The Planning Board of the Town of Ramapo*, 334 N.Y.S. 2d 138, 30 N.Y. 2d 359 (1972).

[12] The case for use of development districts was first made by Marion Clawson, "Suburban Development Districts," *AIP Journal*, May 1960. Also see Henry Bain, *The Development District* (Silver Spring, Md: Maryland–National Capital Park and Planning Commission, 1968); and *Planning, Zoning and Development of Central Business Districts.*

funds from regular operating agencies, and a special tax within its district to allow the provision of special amenities, such as pathways or recreation facilities, or added community services, such as early childhood education centers, manpower development programs, or free public transportation.

Development districts offer a means of circumventing the frustration of uncoordinated public participation in development, a means of properly staging construction activities in well-planned locations, and a means of creating a very attractive market for private investment in locations that make sense in terms of metropolitan objectives.

A New Approach to Metropolitan Planning

Short of the use of development districts, or even in combination with them, the most effective public tool for guiding development is a coordinated capital program. Disciplining the separate public works systems into a unified metropolitan strategy requires major revisions in the role and content of metropolitan planning. Metropolitan planning will have to amplify its function of setting goals, by scheduling and allocating resources to achieve goals. This will require changes in state planning law and involves making metropolitan planning *in these terms* the nerve center of metropolitan policy making. Institutionally, metropolitan planning needs a political home. In the short term, that home may be state government, as part of an expanded state planning process, rather than metropolitan government, although the latter would probably be preferable. Wherever it is located, officials will have to be educated in how to use it to consolidate functional planning and to cast the metropolitan planning system as the entrepreneur of development in the public interest.

The real trick in metropolitan development reform is to create the kind of political environment in which a strategic combination of reforms can be used in concert to make possible the management of urbanization. Unfortunately, the present system is a live and going concern. It is pluralistic, and each element serves well some established interest. A constituency for officials who do not yet exist is needed to counter the political and technical monopolies that now determine development policy. It will be necessary to use the available political processes; ironically, in this situation, the best hope for metropolitan reforms that can really alter land development is often the state, with some federal backing.

2 Fiscal Issues in the Reform of Metropolitan Governance

JULIUS MARGOLIS*

Introduction

Everyone is agreed that state and local governments are facing a severe fiscal crisis and that the structure of local urban governments needs an extensive overhauling. There is no consensus, however, that the absence of some form of metropolitan government contributes to the fiscal problem or that a solution to the local fiscal crisis entails a metropolitan-wide program. The hypothesis of this paper is that the existing structure of metro governance contributes to the fiscal crisis, and that policies to alleviate the fiscal crisis should involve changes in the structure of government as well as in the pattern of local taxation and spending.

The fiscal problems of local government go beyond the pressing needs of financing public services, although the demand for public services and their costs do generate the problems. Inadequate revenues can be, and usually are, a consequence of deeper difficulties. Further, the search for purely fiscal solutions may exacerbate urban problems. Any set of financial proposals will have serious implications for the spatial distribution of programs, the capacity of government to provide the desired quantity and quality of public services, and the distribution of benefits and burdens among socioeconomic groups. Therefore policies for local finance, as well as for local government structure, should be based upon a fairly broad set of social and political goals.

*Director, Fels Center of Government, and Professor of Economics, University of Pennsylvania.

From one viewpoint, it is almost trivial to say that the local fiscal problem is one of government structure. After all, taxes are prices paid for public services, and if the utility of the services is greater than their costs, the benefiting public should be willing to tax themselves. If they do not, then there must be shortcomings in the structure of decision making and in the distribution of government responsibility and authority which prevent the reaching of socially desirable objectives. But structural reorganization will not necessarily correct such imperfections, nor ensure that an optimal public sector, if it could be defined, will be reached. It may be that imperfections in government are basic and at some level irreducible, but we should be able to move toward this irreducible level.

Governments, as we have learned from the extensive literature on public goods, do not function with an individual paying a price, or tax, equal to or in proportion to the values he receives from the public output. Individuals are taxpayers, consumers of public services, and voters, but none of these functions is designed so that the output of the public sector approaches some type of social optimum. Taxpayers' revolts should not be surprising under these conditions. The thrust of my argument, however, is that there are feasible changes in the set of financing instruments and in the structure of government by which public acceptance of improvements in the public sector is likely.

Local fiscal policy includes more than taxation and public services, but much that goes on in the metropolitan area can be classified as government strategy to reduce the tax costs or as individual tactics of moving to avoid costs. Zoning maps, subdivision controls, transportation plans, and a myriad of daily decisions by local governments are all heavily influenced by a concern for their effect on the tax rate. One would more likely be right than wrong if he sought a "tax-rate change" explanation for any randomly chosen city council action. Individuals may not be as alert as public officials to taxes and service, but their behavior is affected and thereby the welfare of the metropolitan area. Tax payments are a heavy cost to be borne by households or business residents, and, so long as there are differentials among the cities of metro, fiscal policies will help determine where households and businesses decide to locate. Introspection, legends, economic theory, and the broadsides of taxpayers' associations tell us that tax and service differentials play significant roles in urban development, though quantitative research has not been able to verify these claims. Despite the absence of empirical confirmation of individual responses to fiscal decisions, it is clear that public officials consider fiscal behavior of great significance for urban development and act accordingly.

The analysis of the fiscal aspects of metropolitan governance opens many questions beyond the choice of tax instruments and levels and distribution of spending. The structure of political decision making, the life style of communities, the pattern of social and economic development, and the creation of ghettos are all determinants and consequences of local fiscal policy. We cannot burden this brief survey with digressions into all the ramifications of local fiscal systems, but we will have to treat many of them, though too lightly.

In the following section, we shall briefly describe some of the characteristics of the current metropolitan fiscal crisis. The next section will develop the principles we should keep in mind when we consider solutions for the problems. The final section will deal with policies. If the argument at times seems tentative, or even diffuse, much of it can be attributed to the unsettled state of the field: there are simply too few firm hypotheses about urban government or local public finance of relevance to metropolitan policy formation.

The Etiology of the Metropolitan Fiscal Crisis

The most common statement of the urban fiscal crisis is that the level of public services is too low or that it is becoming increasingly difficult to raise the funds to support a socially desirable level of local public services. No one would dispute that a higher quality and quantity of public services are to be preferred, but, of course, some would object to the sacrifice in private resources necessary to expand the public sector. I am not interested in judging the relative merits of public and private consumption, but in a related issue — how the structure of government and the tax institutions affect the public services and thus the balance between the private and public sectors. But before dealing with this central issue, let us draw some conclusions from the many studies about state and local fiscal matters.

Aggregate Local Public Services

There is a widespread feeling that there has been a decline in the quantity and quality of urban public services. Do we mean that they have not grown as rapidly as the private sector, or that they have deteriorated in absolute terms, or that there has been a shift in preferences for local public services so that relatively larger amounts have to be spent on the public sectors? Probably all of these have taken place, but every claim cannot be made against each public service or type of government.

Distress about public services usually leads to calls for higher local public expenditures and, of course, revenues. In fact, the aggregate of local public expenditures (smaller than the military budget) has kept pace with the gross national product. Clearly, the issue is not one of an aggregate unwillingness to continue to support the local public sector relative to the private, but the aggregate figure may be deceptive. Expenditures have kept pace, but service may not have, especially if the prices of publicly employed resources have increased relative to private. This has occurred. Therefore, though the public has been willing to support the local public sector in proportion to income growth, the shift in relative costs has meant that a higher percentage of local income must be devoted to the local sector in order to maintain the balance between real public and private output. Furthermore, preferences for public services have changed, and center-city burdens have become very great relative to the suburbs. All these factors have caused the urban public sector to lag behind the private.

Two factors seem to account for the increase in cost differentials between the public and private sectors: (1) a lag in productivity in the public sector and (2) the congestion costs of dense urbanization.

Wages in the public sector increase with wages in the private sector, but productivity in the service industries, including the public sector, has lagged, and therefore the unit cost of public output has risen relative to private. Most public services are very labor-intensive and, unless there are major improvements in delivery systems, their unit costs will be very sensitive to changes in wage rates.

It is argued that our large metropolitan areas have reached the point of increasing costs. At a low level, density means reduced unit costs, but at a high level it causes congestion, resulting in higher resources costs to achieve a constant output level. Though it has been difficult to document the magnitudes of congestion costs, the probability of increasing public costs is very high.

One solution to both sets of determinants of increasing costs (congestion and lag in productivity) is to increase revenues to the dense central cities of the standard metropolitan statistical area (SMSA), but a more basic response would be to improve technology or to redistribute economic activity. Neither would be simple to do, but such efforts would be more relevant than changing government structure or finding financial gimmicks in order to attain the same real levels of services. If the growth of public productivity has been slow relative to private productivity, or if there are significantly increasing costs in dense areas, public services should not grow relative to private services, since the private sector will provide relatively more output per unit of input. One way to offset this shift is to encourage the growth of less dense sectors, where public output is relatively less costly.

The policy implications of the preceding paragraph should be stressed. Men are prone to argue that the center city is the heart of the metropolitan area, that its high public costs should be borne by the entire metropolitan area, which can exist only because of the services provided by the center. This argument, though often accepted, is rarely persuasive enough for action. It is true that the center is more costly, and those who stay in the center do so because of the advantages it offers them. (Even the blacks, who have been discriminated against in movements to the suburbs, have many advantages in center-city locations.) Since the center is costly, it is efficient to encourage outmigration of those who do not benefit and therefore are not prepared to pay the higher costs. Reducing the costs to center-city residents by asking the entire metropolitan area to share the burden masks the high costs of center-city living and improperly distributes persons over space.

It is possible that the gains in private accounts might offset the losses in public accounts and that the dense parts of the metropolitan area might be viable and might become even more dense. Could we redesign our cities *de novo*, we might have areas with even greater density than the existing patterns. At present, however, the dense areas are the older sections with inefficient street patterns, aging utility pipes, and outmoded designs of residential and business structure. The increasing costs associated with congestion are compounded by the nonadaptability of an obsolete private and public physical plant. Increases in productivity and mobility are intimately tied to the thorny problems of renewal of physical plant. The sites that need renewal are the older central cities and railroad suburbs, but the bulk of federal subventions to the metropolitan areas (mortgage insurance, highways, construction of public buildings and facilities) has been directed toward the newer suburbs. Density is associated with physical and spatial obsolescence; public policy has inhibited the "natural," difficult replacement of the obsolete by subsidizing investment in new sites.

Unfortunately, the aging urban structures are associated with low-income populations. The physical plant is more costly to service; its value is low and thus it cannot be taxed heavily. At the same time, the associated low-income residential population, which cannot contribute heavily in taxes, creates relatively heavy public costs. The concentration of physical and spatial obsolescence and special-problem populations results in a relatively heavy tax burden for the well-to-do or new business investment. It is no surprise that the rich and new investment locate in the choice suburbs, even though thereby they increase the burdens for the central city, reduce its effectiveness, and finally give rise to federal or state redistributive programs to salvage the central city and the metropolitan area.

The earlier remark that it may be good policy to let the center city bear its full burden of being more costly seems specious when one recognizes that many center-city residents are restricted to the ghettos and dilapidated districts of the center city. They are not mobile; they cannot readily escape the high costs of the center city. They do not choose the center city for its benefits. The assertion of the immobility of the poor may be questioned, but even if we accept it, we should try to develop policies that aid this group directly, rather than changing relative prices, which inhibits efficient development.

Specific Public Services

Changes in preferences and costs have not been uniform among public services; therefore a single strategy for the public services does not exist. The problems of the public services are not the theme of this paper, but they cannot be ignored, since government responsibility for public services to the metropolitan area is fragmented. Municipalities usually do not finance schools; police and courts are provided by several government levels; transportation needs are not uniform among governments; welfare is often a county or state function; and so on.

Education offers a classic illustration of change in public preferences: a substantial increase in relative expenditures on schooling and still a dissatisfaction about the status of the public service. The percentage of the population in school, the percentage of children finishing high school, or any other index one might consider shows a great increase in educational services. Clearly society has been willing to put relatively more resources into education, yet there is a widespread feeling that the returns are inadequate. There are strong sentiments that federal grants should be increased and that the state should replace the school district as the fiscal agency for schools. Undoubtedly, there is a fiscal problem facing the metropolitan school districts, but it is due more to the effort to transfer a much greater proportion of resources to education than to the fiscal poverty of the metro. Another large part of the school finance crisis stems from the pecuniary problem of increased prices for inputs. The number of teachers has increased at a rate four times that of the labor force. It is not surprising that teachers' wages have also relatively increased.

The educational problem is not uniform over the metro. The complaints of the center-city residents are not those of the suburbanites, nor, of course, do the diverse suburbs have uniform complaints. Similarly, the fiscal solutions are not the same. The booming, lower-middle-income suburb without industry has very different needs from the center city with its large industrial base, relatively large municipal expenditures, and lower-income population with difficult schooling potentials. Grant-in-aid formulas have been addressed to suburban rather than center-city needs.

The increase in the level of education has been primarily among lower-income groups. Relatively more children of the working class are finishing high school and going on to college than in the past. Since these children have less parental investment in their pre- and early school years than children from more affluent backgrounds, public educational costs to bring them to the same achievement level are greater. Unfortunately, the population requiring the greatest public school investment is concentrated in the central city, where the other public costs are also greatest; therefore we find a lower per pupil outlay.

Solutions to the problems of school finance usually take the form of state absorption of costs via fiscal transfers to school districts. Though this will reduce local fiscal burdens, it is not clear that it is optimal. We will postpone discussion of the voucher system to a later section, but even within the sector of public schools, scale economies are lost in many suburban systems, and center-city systems are overly centralized. There are significant cost-per-pupil differentials among systems, but these mask what are probably much greater cost differences per unit of "effective education." A metropolitan plan for education is more desirable than a state plan, since the problem is not simply one of fiscal redistribution but also of program design. The possibility of scale economies, the problems of innovation, the potentials of supplying diverse educational programs, and the advantages of following children as their families move within the metropolitan area all can be handled better within a metropolitan area than within a state system.

Solutions for education finance are usually divorced from the fiscal problems of the community supporting the school district. This failing creates an inequity: the center city, burdened with heavy welfare costs, has relatively less available for other public services. This argument implies that fiscally independent governments are an illusion. Since they must compete with each other for the limited fiscal capacity, there must be a political resolution of their competing demands. Certainly it is difficult to deal with one major service without at the same time affecting the others. Budgetary interdependence among public services is clear, but it is not obvious that creating a unified government to centralize the budgetary process is better than permitting the present political independence. A unified decision-making body is not always the most effective.

Perhaps an even more dramatic shift in preferences is revealed in welfare programs. Despite heavy increases in federal support, the burdens on states, urban counties, and central cities remain very great. Welfare programs are redistributional, and so long as the financing is tied to the territories in which the problem populations live, intolerable strains for those jurisdictions are inevitable. There may be a political consensus to increase support payments to the poor, but those who must bear the burden will move to the adjoining town where there are no poor and escape the costs.

Welfare and education illustrate not only shifts in preferences but also the dilemma facing cities with limited territorial jurisdictions. These services do not benefit all and are therefore redistributional. In the center city, the educational redistribution is to the poor, which aggravates the pressures on the rich to leave the center city. These redistributional activities should be viewed from a metro or even a national perspective. The concentration of the low-income populations in the center city is a service function to the entire metropolitan area; these people, mostly rural immigrants to metro, supply low-cost labor to the area. The investment of resources to upgrade the poor benefits them, but not necessarily the central city, since the poor too often follow the rich to the suburbs when their abilities improve. Labor is very mobile, and a community that invests heavily in its improvement may find that the assets it would have created become located elsewhere. Therefore, there is local resistance to such investment, which is strengthened by the movement of the well-to-do. The territorial fragmentation of the government of the metropolitan area is the source of the fiscal problem of the central cities that house the low-income population, but fiscal rather than structural solutions may offer the best corrective. We will discuss this possibility in a later section.

Distribution of Fiscal Resources

The local fiscal crisis as generated by service levels is clearly, then, not uniform over municipalities or by services. But, the changing preferences for and costs of public services are only part of the picture. The other part of the fiscal problem is the distribution of fiscal resources. State and local expenditures have kept pace with income growth, but local taxation has not. Far from it. Locally raised revenues as a percentage of local expenditures have fallen from 82 to 53 percent over the past sixty years; it has been state and federal grants that have permitted local expenditures to grow. It is reasonable to assume that this institutional change will persist, and increasing reliance on higher levels will undoubtedly be one of the major policy instruments. Grants from the federal or the state government have not been for general support, however, but for specific functions. Since functional needs are not uniformly distributed among the cities of the metro, grants have not been uniformly disbursed.

Unequal distribution of resources among the regions of the metro need not be a problem, however. Every city is composed of sets of neighborhoods; scale and agglomeration economies in the provision of services, as well as preferences for like-minded neighbors, create these divergences. It is only natural that the neighborhoods reflect different preferences for public services and different allocations of resources to support these services. Outside the central city, however, neighborhoods frequently become municipal governments, and neighborhood differences thus, unfortunately, become institutionalized inequalities among cities, which seek to perpetuate this inequality.

Table 1 illustrates the disparities in one metro, Milwaukee. The first two columns show the unequal distribution of fiscal resources among the types of cities. The central city is not always the poorest of cities—there is often more than one poorer low-income residential city—but the ranking of the central city below many of the types of suburban communities is quite common. Typically, the central city's municipal expenditures are the highest in metro, although in Milwaukee the rich residential suburbs spend more for municipal

TABLE 1. Fiscal Disparities in Milwaukee SMSA, 1966

Type of city	Property values per capita (1)	Income per tax return (2)	Municipal expenditures per capita (3)	School expenditures per capita (4)	Mean property tax rates (*mills*) (5)	Benefit-effort ratio (6)=(3 + 4) / (5)
Central city	$ 5,454	$ 4,697	$ 81.3	$ 70	40.7	3.71
Balanced	8,982	5,532	62.5	103	29.5	5.60
Industrial	25,239	5,112	142.9	104	24.3	10.16
Residential High-income	10,937	12,273	118.5	149	29.4	9.08
Medium-income	7,180	6,642	54.3	121	28.4	6.17
Low-income	5,799	5,436	47.9	115	27.9	5.84
Urban area suburb	9,219	7,197	72.5	121	28.4	6.82

Source: Data from John Riew, "State Aid for Public Schools and Metropolitan Finance," *Land Economics* (August 1970).

purposes. The per capita school expenditures are greatest in the suburbs, reflecting a lower level of services and a disproportionately smaller school-age population in the central city. The fifth column shows the local tax effort. The last column gives the benefit-effort ratio, which shows the dollars of expenditures per unit of mill-rate. In rough terms, it shows the quantity (on the assumption that costs of public services are reasonably similar among the cities of a metropolitan area) of public services supplied per percent of family income devoted to local taxes. In the Milwaukee area, a taxpayer would receive two and a half times as many public services for the same 1 percent of income given in taxes if he lived in a high-income suburb as he would in a central city. He would receive one and a half times as many services if he lived in a high-income suburb as against a low-income suburb.

The differences in residents' benefits per unit of effort is really much greater than reflected in this table. Much of the expenditure of the central, the balanced, and the industrial cities is directed toward servicing the business activity in the community; therefore the expenditures overstate the residential outputs relative to the residential services in the residential cities. Further, the central city is the repository of the special-problem populations, who are more costly to service, so still less is left to provide "normal" residential services.

The major factor accounting for the great difference in output per unit of effort, or its inverse, tax price, is the wealth of neighbors. If the cost of providing a unit of police service is the same in two cities but the average income of one city is twice that of the other, then the percentage of income paid to hire the policeman in Poortown will be twice that of Richtown.

The initial response to the disparities revealed by the differentials in benefit-effort ratios is to deplore the inequity among communities—we could call it horizontal inequity. Perhaps more important for urban policy are the incentives created by the existence of horizontal inequities. Clearly, everyone in a municipality gains if a new resident has an income or property value greater than the average and generates a public service usage less than the average. Therefore fiscal profitability becomes one of the major goals in shaping land-use programs in a city. Unfortunately, these programs are of the beggar-thy-neighbor sort and can lead to a decline in the aggregate welfare of the metro. It is not simply that city A's gain is city B's loss; the process of trying to shift the location of fiscal gainers and losers means that individuals will not be located at sites perhaps preferable by all other criteria. That is, a poor person employed in a plant located in a suburb may be unable, because of zoning restrictions, to find housing in nearby suburbs. Consequently his and everyone's travel costs may increase (additional highway capacity may be needed, for instance), or the employing firm may have to pay higher wages with a consequent incentive to relocate. In any case, nonoptimal location will result, and the low-income, tax-burdensome resident will continue to impose costs greater than the tax revenues he generates, wherever he settles.

Two other aspects of these perverse fiscal incentives are of interest both as sources of problems and as indications of directions for policy. If the rich succeed in achieving homogeneous suburbs of the rich, then there will be a decline in the magnitude of redistribution that occurs through the fiscal system at the local level. Redistribution has long been accepted as a public function and, though there are severe restrictions on how much redistribution can be undertaken by local governments, some of it is. Education, hospitals, and welfare, to name a few local services, have varying degrees of redistributive effect. In general, households pay local taxes as a constant percentage of income and, though public services are not distributed equally, there are strong pressures for uniform public services; therefore the rich support the poor in center cities. Certainly, the poorest in metro do not contribute sums equal to the services they receive. And the greater the number of rich who leave the fiscal base that supports these services, the greater the burdens imposed on the well-to-do who remain, or the poorer become the services. The dynamics of the situation suggest that the well-to-do will have still stronger incentives to leave, and the services will become still poorer.

The moral of the above is that redistributive financing will have to be shifted to higher levels of government, which have jurisdiction over the well-to-do as well as the poor. In the long run, the rich will not escape the burden of support of the poor, but the process of "search-and-tax" is a painful one. In practice, it has required a shift of financing of redistributive activities to state and federal governments. As we noted above, the percentage of local expenditures financed by locally raised revenues has declined dramatically. The current agony of center cities and states makes it clear that pressures to increase the proportions of intergovernmental transfers will continue. Though most redistribution expenditures should be borne by the federal government, a metropolitan government could play a useful role in intrametropolitan transfers, especially where the implicit redistributive aspects of normal public goods are concerned.

Redistribution can be considered as one case of a general class of externalities—i.e., interdependencies in production or consumption, but where the affected parties do not have an exchange relationship. These interdependencies are complicated by the existence of many small governments and exacerbated by the socioeconomic differences among the communities. Income stratification is only one of the differences among the communities. Some are industrial satellites, some are middle-class bedroom communities, some are ethnically homogeneous, and so on. Whatever the source of differentiation among municipalities and homogeneity within communities, we would anticipate that the suburban homogenization would lead to an increasing amount of activity across communities. That is, a larger proportion of the population would be splitting their working, shopping, sleeping, and so on among more than one city. Although this has some attractive features, it leaves the individual without a government concerned with the total set of activities of importance to him. Split responsibility would be no problem, were it not for the nature of the interdependency among the communities of a metropolitan area. Automobiles may come in family units, but a highway network has large lumpy elements that cannot be decentralized to cities. We know that cooperation among the cities of a metro is poor; we do not know the extent of the losses associated with uncooperative behavior. Fiscal considerations often play a role in the debates about possible cooperation; how decisive they are in the final outcome is not known.

Horizontal inequity, distributional shortcomings, and allocational deficiencies are all part of the structure of fiscally independent local governments seeking to maximize the difference between the benefits they receive and the costs they must bear. Against these losses should be placed the virtues of a more homogeneous community. If neighbors share preferences about public services, it is easier to get consensus about the quantity and quality of those services. There will be less controversy over a playing field versus a science lab, a park versus a library. Not only will the decision-making costs be less, but there is a greater probability that the outputs will be more satisfactory for most of the residents. A set of public services that matches the preferences of the residents is no mean accomplishment. On the contrary, a whole new school of thought has developed which elevates these gains to the level of dominant objectives. We do not know how important it is to have significant differences in packages of public services. In principle, it should be very important, but in practice there are huge differences in the package of public services within the boundaries of a single large city. No one knows whether the differences among the neighborhoods of a center city are greater or less than the differences among the suburban cities surrounding the center city. However, the tax rate within the center city is uniform, and it is likely that the resident of the suburb feels that the political leadership is more responsive to his complaints, though in the latter instance again we have no evidence about the facts.

One might conclude that a world of homogeneous communities is likely to be without "fiscal problems," since there will be local consensus about the quantity and quality of public services. But this will not necessarily ensure the proper quantity and quality of public services for the metro, since the existence of externalities may mean that local optima for each city, taken one at a time, still leave an unsatisfactory state for the metro. However, it might mean an absence of distress and concern about the ability to raise funds to cope with local problems. Since we do have a large sample of suburbs with varying degrees of homogeneity, this hypothesis should be capable of testing. We have touched on a variety of questions concerning the aggregate level of public services, some specific services, and fiscal capacity differences, and the more plaguing issues raised by a plethora of governments, divided by territory and function. We have looked briefly at many of the problems repeatedly stressed in the literature. But before turning to the policies likely to overcome some of the metro fiscal difficulties, let us briefly outline some analytical categories implicit in the preceding discussion and useful for the consideration of policy.

Fiscal Principles for Metropolitan Reorganization

It is traditional to view local public finance from the single perspective of taxes necessary to finance needed levels of public service. This view is inadequate if we are concerned with the optimal structure of government, its provision of services, and the distribution of benefits and costs. To focus on optimal finance and governance, we should adopt the perspective of a metropolitan public economy. To do this, we should look at the local public sector from a series of viewpoints. Different levels of analysis are desirable because decisions can be partitioned, and not all public services require the same government structures or financing arrangements. Partitioning of decisions can be facilitated by fiscal policies; financial instruments can provide incentives and constraints for public decisions to harmonize decentralized decisions with an agreed-upon set of social objectives.

Metropolitan Governance and Federal Fiscal Structure

To support an active amount of commercial and social interchange among the many metropolitan areas, there has to be a reasonably uniform body of rules and government practices. U.S. history, as well as that of all other industrializing nations, reveals the growth of centralized authority. Federal authority extends into production processes (the post office and nationally regulated telephone system create a uniform communication system), rules (federal courts constantly limit the authority of state and local governments), consumption (grants for open spaces as well as for many other functions try to create uniform public services), and redistribution (welfare programs are basically federally imposed). The power and influence of the federal government is recognized, but, since its role in metropolitan governance has never been analyzed, it is difficult to specify its characteristics. The current revenue-sharing debate deals implicitly with the optimal division of authority and responsibility in a federal system, but the level of debate on these issues is very low. We shall not elevate the policy discussions with any general analysis, but there are a few points where the role of the federal government in the fiscal structure is relevant and touches on the intrametropolitan fiscal and governmental structures.

It has become traditional to divide the fiscal functions into three categories: stabilization, redistribution, and allocation. The first, stabilization, is always restricted to the national government, since it has the monetary powers of the central bank. Only the federal government can incur debts without fear, and, though there is a problem of integrating state and local fiscal actions with the federal government as it pursues a stabilization policy, this has no bearing on the structure of metropolitan governance. It might be easier for the federal authorities to integrate state and local fiscal behavior if these governments were fewer in number, but this seems a weak argument for metropolitan government.

The second function, redistribution, has more significant implications for metropolitan governance. In general, redistribution is assigned to the federal government because of its superior ability in using a progressive income tax. Any state or locality that tried to approximate the scaled levies of the federal income tax would soon discover that the upper incomes would move to other jurisdictions. Assigning redistribution functions to the national level is sensible; unfortunately, the state and local fiscal structures have significant redistributive aspects that have serious intrametropolitan consequences. We shall analyze these in terms of the allocational function, since they are usually by-products of efficiency objectives.

The allocational function deals with the provision of public services. These are considered appropriate for local government, but, as we shall see, they pose great difficulties for our fragmented metropolitan area. Public services can be divided into two groups: public and merit goods. Public goods are characterized by externalities in production or consumption: a water supply system is usually public and exhibits extensive scale economies, with great inefficiencies if more than one system supplies the same area. Such a producer is not always a public agency—a regulated private monopoly may be as effective—but it is always publicly controlled. Externalities in consumption take a different form. For these goods it is technically impossible or very costly to exclude consumers. For instance, a reduction in air pollution cannot be denied to any resident.

Merit goods are more interesting to us than public goods. Merit goods are usually defined as private goods; that is, there are no significant externalities in production and excludability is feasible, but households are as a matter of policy encouraged to consume more of the good. An illustration would be schools. The technical character of a merit good is that the index of well-being of one individual increases if another individual consumes more of that good.

In the case of merit goods, we would expect that individuals would support the government in underwriting the provision of some goods, that is, that citizens would pay taxes greater than the benefits they receive directly so that others could consume more of those goods. The problem in urban areas is that almost all goods are treated as merit goods, whether they are or not. A merit good calls for some order of redistribution—there is an increase in income-in-kind to those who are to consume merit goods. If we are correct in asserting that local public services are treated like merit goods, then we have a significant level of redistribution taking place at the local level. This has the harmful effect we mentioned earlier: those who lose flee the redistribution.

The rationale for the treatment of local public services as merit goods is political. The tax payments to the local sector are roughly proportional to income. In preceding decades public services were distributed in proportion to income, but this has been changing. The political demands of the poor have grown and are likely to accelerate. A principle of uniform treatment has always held for government; though it has been violated on a grand scale in the past, it has justified the political response to the demands of the poor. But, as we have seen, the upshot of proportional taxation and equal services is that the rich, who do not receive benefits in proportion to costs to them, will abandon the cities of the poor for the suburbs of a politically fragmented metropolitan area.

Redistribution, widespread in the local sector, therefore places a great burden on the design of a local government structure. The "natural" response—suburban fragmentation—poses problems for the provision of public goods, which is supposed to be an activity peculiarly suited to local governments.

The provision of public goods can be viewed as a case of natural local monopolies or of group purchasing. It would be absurd to have houses hook up to competitive water mains and extremely costly to light a street house by house. The first case, production economies, usually calls for larger jurisdictions; the second case, group purchasing, usually implies neighborhood controls. Combine these two technical characteristics of public goods with the redistributive aspects of public services, described as the merit good case, and the complexity of the problem of metropolitan organization begins to take shape. To these features we must add the even less understood problems of political decision making. It is inadequate to view the public sector as a set of services to be supplied one good at a time or to one area at a time; for each function or area we have a constituency of citizens, affected parties, elected officials, and a bureaucracy. We have a complex set of political processes, and an "economic" solution for any part of the public sector is very likely to founder on the political interdependencies among the set of activities.

We have briefly discussed the local public sector from the perspective of functions. Now let us look at the same sector from the perspective of organizational analysis, before moving on to a consideration of policy.

The Distribution of Governmental Roles

A government can be viewed as an agency wielding power to affect social policy, with far-reaching authority and concern for all problems. The reality of local government, however, presents a very different picture, though not necessarily a less viable one. Because the metropolitan area is highly interdependent, there is a strong presumption that some form of metropolitan governance is needed. But even those who give great weight to government as a shaper of social policy may find a unitary metro government neither necessary nor optimal. The appropriate distribution of authority and the assignment of fiscal instruments should be based upon a complex consideration of production relationships, political responsibility, and social goals. Some elements of governance can be delegated to territorially or functionally constrained units; others must be retained by a unit with wide authority. Let us briefly sketch out some of the considerations.

The partial government. There are defensible arguments for the establishment of limited governments, either with full fiscal independence, or so structured as to be almost independent of their local formal authority. The questions are: which functions should be subdivided and how, how should they be financed, and how is finance related to the likelihood that these limited governments will operate wisely. It is an empirical fact that public participation in the policy making of the limited governments is meager. When such governments are run by elected officials, elections are often uncontested or the vote is light. If supervisory boards are appointed, they are often anonymous as far as the community is concerned. Direct popular checks on these governments are weak. The political process, which might provide an audit of productivity and responsiveness to citizens' preferences, is carried on quietly, away from the public eye. The one exception is the school district, where the local tax burden is great and the quality of service sufficiently important to a large enough percentage of the electorate that there is often vigorous interchange between the citizenry and the limited government.

The great bulk of local decisions are made within a very restricted framework. Consider the school board. In principle, the siting of schools should be related to residential location plans and transportation design, and curriculum should be related to the metropolitan labor market, but the major concern of the school board is with issues only marginally related to what goes on elsewhere in the public or private sector. Though the local school board is heavily affected by metropolitan activities, it cannot be expected to have much concern for the state of the metropolitan area. Activities of functional units like the school boards can be divided into two categories: production and demand.

Production efficiency is associated with managerial capacity to do as well as one can in producing a given package of outputs and in achieving the scale of operations so that the cost per unit of output is minimized. Responsiveness to demand is linked to production in that we must determine the desirable scale of services and assign a value to the units of output. A mass transportation system can be designed to carry passengers with a smaller per unit cost than a highway system, but the adoption and design of mass transportation should be a function of the values assigned by the potential users to highway versus rail carriers.

In the local public sector, governments limited to production and demand conditions for a specific service in a limited area are common. Many local governments have been established for a single purpose—in some cases with a single tax base, in others relying on user charges. The more common situation is a multipurpose government, but even here we find special boards, commissions, clientele groups, entrenched bureaucracies, and similar structures, so that some local governments can be viewed as coalitions of special-purpose agencies.

The general government. Though most local decisions are restricted to narrowly defined problems, the big decisions impinge on the total society and therefore some institutional concern for the whole is necessary. The highway department can destroy the central city; the school or police department's behavior may strike the spark that ignites the ghetto, and so on. The extreme interdependence does not necessarily mean that there has to be a unified plan, however. After all, the economy is the most complex interdependent structure we have, and it operates tolerably well with only a few, though extremely important, overall control mechanisms. The fact of interdependence does not imply that central administration is necessary, but it does lead to a concern with a broad set of consequences, and, therefore, to an involvement of many factions in the political bargaining arena.

For example, the diverse policies of local governments regarding taxes and expenditures loom large in the locational and political participation decisions of individuals. A simple shift from free water to average cost or marginal cost pricing of water will alter development decisions. A marginal cost pricing of parking space can alter traffic and commercial development. The differences in taxes and services among suburban cities allow individuals greater options for matching their preferences, so diversity would seem to be desirable, but, if linked with a poor fiscal structure, it may have very undesirable effects for the functioning of the metropolitan area.

The cases of interdependency that give rise to government actions are called externalities. An externality exists where the decentralized prices associated with activities do not account for all of the benefits or costs. Negotiation through an administrative or political structure is relied upon to correct for this failure. Though the concept was developed to analyze market behavior of firms and households, it is equally valid for the interactions among governments and bureaus. In fact, it is an even more central problem for governments, since institutional change to internalize externalities (primarily to capture unpriced benefits) is much easier in private firms than among governments. The beekeeper can easily buy the apple orchard, but the police department cannot merge with the courts or welfare agencies.

In the local public sector, externalities occur at two levels. The most discussed case of local public externality is intermunicipal spillover. Activities in city *A* generate costs or benefits in city *B*. If city *A* is not assessed a penalty or not rewarded for these "exported" effects, an improper amount of production will take place in city *A*. The fragmented jurisdictions of the metropolitan area guarantee that these externalities will arise, and the absence of any metropolitan government makes it very difficult to overcome them. The proposed fiscal solution to this problem takes two forms. The common recommendation is a categorical grant to encourage the exporting city to increase the supply of services. A more debated solution is the commuter tax, which seeks compensation for the services supplied to nonresidents. Both solutions are special cases of beneficiary taxation, or user taxes. We shall discuss these further when we deal with policies.

The intermunicipal externalities have been growing with the flight to suburbs. They become more intense as the suburbs become more homogeneous internally, a process we described previously.

A second source of externalities lies in the budgetary process of the local government or set of governments. It is clear that the zoning board affects schools, which affects parks, which affects police, and so on. None of these agencies has direct market relationships, and therefore gains or losses to the community due to interrelations among these functions are difficult to affect, except by negotiations within a single responsible government or among sets of governments.

Were it feasible to develop market relationships among services (a park district renting or purchasing space from a school district, for example), there would be a stronger case for separate functional governments. But even then we would have to question the effectiveness of an electoral process for each function as a means to create representation for the citizenry. Without strong representation, there would be no reason to assume that the commercial contracts among functions would be in the public interest, even if they were feasible.

If we gather the agencies into a single government structure, have we solved the problem? As the scope of decisions controlled by the government grows, there is likely to be more community interest in the electoral process, but the interpretation of the vote in terms of welfare becomes very clouded. A vote for a representative becomes a vote for a very complex set of public goods that few citizens will have evaluated. A voter will be asked to pay one tax, but there is very little likelihood that he can cast a vote so that his legislator knows how to balance his valuations of the different public services and have them add up to at least as much as his tax. This messiness of representation is reduced by territorial governments of homogeneous population with similar tastes. But this solution creates problems of scale economies, intramunicipal externalities, and distributional shortcomings.

The above conflicts alert us to the certainty that there is no simple answer to metropolitan governance and little likelihood of an optimal solution. But it would be even more depressing if we knew an optimal solution and did not know how to get it adopted. We shall use all of these considerations, in the next section, as we discuss policies.

Opportunities for Metropolitan Fiscal Reform

Intergovernmental Fiscal and Functional Transfers

The historical trend toward centralization of both service and taxing function is very clear. The centralization of services has been vigorously criticized, but to little avail. The current effort to reverse the trend toward centralization of function is to increase centralization of taxes, which are to be transferred to local governments for discretionary spending. As expected, this is a policy popular with mayors and governors, though if revenue sharing should fail, local leaders would certainly support the further transfer of additional functions in order to ease their fiscal problems—i.e., more centralization. The major functional transfer proposed is the funding of welfare by the federal government and of education by the state government. Certainly, if both these transfers were to be effected, and the trend is strongly supportive of these transfers, then a large part of the fiscal crisis of local government would be solved.

The choice among transfers of functions, categorical grants, or revenue-sharing schemes goes beyond purely fiscal considerations. All of them are discussed in terms of keeping intact the current structure of government, though all of them will seriously affect that structure. Unfortunately, these proposals are usually debated in terms of the fiscal crisis, and the implications for governance are too often ignored.

Federal revenue sharing. Revenue sharing, in terms of our previous discussion, is a mechanism for partially overcoming the dilemma of local finance—tax competitiveness among communities and nonbenefit taxation within a community. The federal tax is uniform over the area within which resources can move, and there is therefore no escaping the burden of taxation. Since service levels are not uniform, however, incentives to relocate are still present. Revenue sharing can be designed to reduce one of the harmful cumulative effects of this mobility: if the city's fiscal policy is redistributive, the out-migration of wealth will not impoverish the community and make the program self-defeating.

Revenue sharing or unencumbered block grants to localities are opposed on two major grounds: (1) the localities will treat these resources as free funds and therefore waste them, and (2) the localities are so poorly managed that they will use them ineffectively without the help of federal guidelines and directives. Neither argument is very persuasive.

These arguments assume that the local government assigns a low value to funds received from the superior government, since there is a long history of local spending of federal grants on projects with low utility while pressing needs go unmet. This sad history derives from the fact that the grants are not transferable among functions by the local government, and therefore nothing is lost by using them, even when the payoff approaches zero. In fact, given the gains that could be made from construction contracts, it might even be advantageous for the locals to undertake the project at a negative return. The local spending of federal or state grants will always increase local incomes more than local taxes paid. It would be hazardous to conclude from this experience that transfer of free funds would have the same consequence. It is not reasonable to assume that the locals would waste the funds were they free to use them for *any* local purpose.

A more persuasive argument can be made about the asserted incompetence of small cities, and even many large cities, to adopt optimal policies and administer them effectively. Here the issue involves professionalization of staff, which is closely related to scale economies. Professionalization in an administrative organization is a function of specialization, which again is a function of size. It may be that the local services gain in citizen responsiveness as they lose in administrative competency, and that we should therefore pay little heed to the argument of nonprofessionalization. One might compare state welfare programs with county programs, for instance, to test their relative virtues. State prisons are better "managed" than municipal jails, but does anyone know whether either are effective in reducing crime? No matter what one concludes about the relative merits of distant professionalization versus neighborly incompetence, one might well prefer neighborly competence. Its achievement, however, is not on the short-term agenda. It is more likely that we will have to design programs allowing for trade-offs between professional skills and community responsiveness. Revenue-sharing schemes, for example, might have built-in requirements for improved administration.

At the minimum, accounting reforms could be required. Program planning and budgeting systems (PPBS) is a much maligned venture today, but imposing large elements of reporting through such schemes can go a long way toward improving competence. A more sweeping reform would extend to the identification of the potentialities of vertical and horizontal integration, which would permit specialization in some processes and community control over others.

A more substantial charge against local competence is the inability of local governments to overcome political inhibitions against government reorganization. The federal record is not much better. Requiring improvements in local government as a condition of revenue sharing rather than retaining an ineffective local structure and a clumsy federal structure is the indicated solution. Federal revenue sharing offers an opportunity to create a new federalist constitution by legislation. The ground rules of the opening debate have been confused, and it may be a decade before the final structure emerges, but it is a signal opportunity to innovate in all dimensions of federal-local relations, as well as in relations among the locals.

Metropolitan revenue sharing. If metropolitan government is not achieved, and it is not likely to be universally adopted in the near future, then a set of governments will remain, with the fiscal disparities depicted in table 1. If it is reasonable to argue that the tax-price an individual pays should not be affected by the wealth of his neighbors, then a fiscal redistribution scheme can be adopted within the metropolitan area that would reduce tax competitiveness and enable cities to carry on their redistributive functions with minimum risk. Essentially the model would attempt to equalize the output per percent of taxable income throughout the metropolitan area, taking into account cost differentials due to urban form and to the existence of special populations.

The argument for tax-price or benefit-effort equalization is that the percentage of income to be paid for public services should be a function of the real resource cost of providing that service and should not be affected by the wealth of neighbors. If achieved, it would lead individuals and firms to choose locations on the basis of real resource costs and output differentials rather than the pecuniary effects of the wealth of neighbors. Of course, individuals may prefer to locate close to wealthier persons for reasons of snobbery or possible commercial connections. A metropolitan fiscal redistribution scheme seeks only to redress the inequity of differentials caused by wealth of neighbors, and to reduce the incentives to locate or to make municipal land-use decisions because of these effects.

The criterion of benefit-effort equalization bears a rough similarity to the traditional economic rule that an individual should pay a price for a commodity equal to its marginal cost. In the case of municipal services, prices (or tax-prices, which are percent of income over output) are a constant mill-rate times the value of real property, or roughly proportional to income. Therefore, if benefit-effort ratios are equalized, an individual's tax payments will vary with his income and with the real cost of providing municipal services, but not with the average fiscal capacity of the municipality.

If we accept this criterion as binding, the metro should assess a sales or income tax on its residents and distribute the proceeds among the municipalities in inverse proportion to the fiscal capacity of the municipality. If the production costs are the same for all communities, the rule would call for equalizing benefit-effort ratios (output per percent of fiscal capacity) for all municipalities. The operating measure derived from this rule is that the metro fiscal transfer should be equal to (one minus fiscal capacity of the community/fiscal capacity of the community with the highest B/E ratio) times budgeted expenditures of the community. This formula is identical to the formula developed by school finance reformers under the title, "percentage equalization."

There are many attractive features to such efforts toward equalizing resources within a metropolitan area, among them more equitable distribution of resources and removal of perverse fiscal incentives for segregation. But, there are still significant shortcomings to be overcome by research.

A metropolitan fiscal redistribution scheme may solve the fiscal crisis of cities with poorer populations, but, like federal revenue-sharing programs, it does not address itself to the underlying cause—arbitrary geographical boundaries imposed on a complex, interdependent economy to create governments whose problems are further aggravated by poor fiscal policies.

Fiscal redistribution can be a vehicle of structural reform. Grants can be refused to governments that are too small, in order to force consolidation. (This was carried out by some states in their school consolidation programs.) Experiments might be undertaken. For instance, some of the state support for education might go to quasi-public corporations that would carry out educational programs. Innovations in programs as well as in government organization might be attempted. State legislation dealing with percentage equalization, or some variant, might define the community for equalization as the metro area in order to buttress the plans for "optimal" development of metro.

Unfortunately, there is no consensus about the optimal government structure. However, some of the principles discussed earlier can guide us in research and in policy development. The current philosophy of proportionate taxation, uniform provision of services, and multiple jurisdictions is self-defeating. Out of it will emerge a segregated set of cities, differentiated by income and by preferences for public services. For the upper incomes, this result is not unwelcome. The losses will be in scale economies in public facilities, irrational land use for metro development, externalities among municipalities, a socially inferior pattern of segregation, difficulties in developing a political consensus on metro problems, and an increase in federal and state programs to overcome these difficulties. In the final sections we will discuss financing policies that may reduce some of the metro problems, but let us first complete the discussion of intergovernmental transfers.

Categorical grant programs. Categorical grant programs have been the major federal response to the metropolitan fiscal dilemma. At first support was directed toward the newly developing suburbs, which were planted in the rural hinterland without a tax base. Federal subventions in new housing and construction of new public facilities made possible the exodus from the center city. The next set of policies was addressed to relieving the burdens of the older cities as they sought to cope with older physical capital and problem residents. Both policies were redistributional (i.e., they served special-population categories), but the former provided large speculative gains for land developers, whereas the latter eased the strain on the cities created by the demands of the lower-income groups for a larger share of the public resources. The differences among categorical grant programs are important. The implications of replacing them by general or special revenue-sharing schemes will not be the same for all programs. Further, federal support might be appropriate for some programs, whereas others might be more sensibly supported from state or even metro levels.

The highway, water, and sewer programs, which create urban accessibility to the rural peripheries of the cities, have benefits restricted to the metropolitan area and concentrated in the suburbs. The gains are localized, and, if a political mechanism could be devised, the metropolitan area or its county should be prepared to finance the program. If the investment in infrastructure is sensible, the gains are greater than the costs, and land values should increase, even after the capitalization of property taxes to pay for the investments. Infrastructure investment is profitable to the local establishment, and it will be undertaken if there is a government with jurisdiction over the territory where the land-value gains would occur. Furthermore, if grants to suburban governments were unencumbered, in the style of revenue sharing, much of the proceeds would go for these same programs. Since many of these programs involve scale economies (water reservoirs, sewage treatment plants, or freeways), grants to suburbs are ineffective. A "regional" agency should design and implement those parts of the programs which exhibit scale economies. Loosely organized regional councils of government will prove ineffective, since conflicts among the cities over the distribution of gains make collaboration difficult. The typical solution has been a special district or a state agency. But neither of these is likely to concern itself with the interdependency among the programs or with the implications of the set of programs for the development of the area. In principle, the state agency is responsible to a legislature, which theoretically serves as a unifying body to consider all infrastructure investment in relation to the development of the area; in practice, the legislative forum is too distant from the scene to coordinate the various regional offices of a state agency.

The second major group of categorical aids is directed toward human investment. Though these expenditures may also raise property values, such gains are small by comparison with those due to infrastructure expenditures. These limited increases in land values are not likely to provide local incentives or sources of fiscal support. A social work agency is unlikely to be a source of capital gains. Better schools will result in higher property values, but the gains for the students are far more significant. Investment in humans is investment in the most mobile of resources. Furthermore, a citizen who does not benefit from such investment is not going to tax himself so that others may benefit. (A citizen may support a freeway connection, even if he does not use it, if he believes it will lead to positive gains in land values.) Categorical grants can provide local services that do not return benefits to all taxpayers. If these were replaced by revenue-sharing programs, the public services would probably revert to a pattern of benefit creation approximating the pattern of tax payments, and human investment programs, especially for the low-income population, would be sharply reduced. This would greatly increase the political tension in the center city, with still further differentiation from the suburbs. The probable consequence would be an increase in direct federal support for the services. Welfare is likely to be the first candidate, but health and education are not far behind.

User Charges and Intergovernmental Contracts

Once we reject a simple solution like a single government for the metropolitan area, we introduce problems of relations among the different governments of the area. These problems are of two sorts: agreements among the governments and the treatment of individuals who move among the governments. The movement may be diurnal, like a shopping or job trip, or it may involve a longer cycle, like a residential move. The concept of user charges or contracts has special merit for metropolitan government because mobility reduces the power of coercion that underlies a tax.

Economists are prone to urge user charges under almost any condition. A user charge is a price for a service, and its voluntary nature means that an individual votes with his purse for the exact amount of the public service he prefers. The objections are that it is inequitable, since the rich will get more; that it is inefficient, since many externality benefits cannot be captured in a price to a user; that it is impractical, since the public services are characterized by a large gap between marginal and average costs and marginal cost pricing thus would not reflect the cost of providing the service; and that it is just not technically feasible for most public services. This is an impressive set of objections, and, though there is a reasonable answer to each of them, we shall not try to develop the case. It should be clear that we have barely begun to exploit the possibility of user charges and that they should be adopted if the objections can be overcome. For instance, few would protest the municipal library's proposal to develop a reference room for investors only if the users would cover the incremental cost of providing the service. Arguments for user charges go beyond the more efficient functioning within a city; they have merit in regard to the complaint of suburban exploitation of the center city and the incentives of the unbenefited to move.

The more the city relies on a user charge, the less concerned it is with the transients' use of its facilities. If they pay the cost, why fret? If the central city can charge for its zoo, it should not object to the suburban family using it. On the contrary, it may gain financially, since all users will contribute, not just the center-city residents.

The incentives to move may be more important. On the whole, the demand for public services grows with income. The upper incomes are willing to spend more for public services, but not if they receive the same amount as the lower incomes, who may be contributing far less. It is probable that higher-income groups can be discriminated against up to a certain point, beyond which they will move to a community with a relatively smaller low-income population and therefore less discrimination. User charges reduce the level of discrimination in a selective fashion and therefore can reduce the fiscal incentives to move. In general, one would opt for proportionate or progressive taxation by the metro or state fiscal authority, together with increased user charges by the municipality.

Contracting among governments is an extension of user charges. Scale economies in some public facilities make it inefficient for a small city to provide a fully integrated set of services. However, a growth in size of the government may make the city extremely heterogeneous, creating all of the difficulties of decision making and allocations of taxes and services among residents. User charges are one mechanism by which the larger city could satisfy the diverse citizenry. Alternatively, smaller governing units could contract with a higher level government or with one of the local governments, for that part of the service which had scale economies. In this case, the municipality would be a club of like-minded citizens sharing a similar view of the public sector, having their preferred quantity and quality of public services, and still being able to achieve the scale economies of large governments. For instance, a small city might contract with the county for engineering design for its streets and sewer or for an information retrieval system for its police.

Contracts need not be between city and county or metro; they may be among smaller cities. The central city often sells services, especially water supply or waste disposal, to suburbs, but these exchanges can also be feasible among sets of suburbs. The scale economies of a fire-fighting force are not great, but usually they imply a service area greater than a suburb. The major gains of a larger area can be achieved by cooperation on operations, but there are still further gains on specialization that entail purchase contracts. The infrequency of contracts among suburbs, however, implies that one should not put too much hope into this pattern.

User charges and intergovernmental contracting further the objectives of scale economies—group purchasing of services—but they are less effective for redistributional objectives and may be counterproductive for integrated planning. No single policy, however, will be effective for all dimensions that should be considered in government design. The best we can hope for is that the set of policies will be able to achieve the desired objectives.

Voucher Systems and Assigned Taxes

A further variant of quasi-market mechanisms that could be used in the metropolitan areas is the voucher system. This system has been analyzed in terms of the use of private supply as an alternative to public production, but may be equally interesting in the case of alternative public suppliers. For instance, the educational voucher could be made payable to an alternative school district as well as to a private school. Or one city might pay a fee to another, based upon the number of its residents who use the other's facilities (visits to parks, treatment in hospitals, or police protection, for example). Essentially, the use of vouchers or user fees increases the flexibility of government organization and creates incentives to alter government in response to demand and production conditions. For instance, if city A computes its marginal cost per pupil and agrees to pay this amount to any other city if its resident goes to school there, there will be changes in the use of schools among cities in response to differences in quality and cost of education.

A logical extension of user charges and user fees is the assignment of a tax to a local government of choice. For instance, the school tax, usually computed separately, might be assigned to any government or even to a private system. This freedom might be extended to the general taxes paid by an individual, allowing him to allocate his tax among municipalities, county, state, school district, metropolitan council, and so on. A scheme of this sort would still further strengthen the set of incentives for improvement in the behavior of government.

All of these quasi-market mechanisms—user fees, vouchers, and assigned taxes—have the virtues we usually associate with the marketplace and, of course, the disadvantages. Such market failures as externalities and distributional effects would be as evident in the public sector as they are in the private. In principle, it should be easier to overcome these imperfections as they arise in the public, quasi-market sector; in practice, however, the public bureaucracies may prove to be as resistant as private firms.

If the quasi-market mechanisms suggested above were introduced, there would be less urgent need for structural reforms. They carry with them incentives for changes in government in response to "impersonal forces" of the market.

Property Taxation

Property tax payments are about equally divided between business and households. As a business tax, property taxation has many shortcomings due to the great difficulties of assessment, tax competitiveness among cities, and the very unequal spatial distribution of business property. As a residential tax, it is less defective, but, since it is a tax on improvements, it discourages investment in the upgrading of central-city housing.

The property tax as a business tax is singularly deficient. Market values are the standard for property tax assessment, but the market in industrial property is inactive and so deeply intertwined with the profitability of existing users that it is very difficult to establish uniform values. Assessments are made on the basis of proxies of market values, which are subject to political negotiation. It is not surprising that great inequities among firms develop. More germane to our interest in metropolitan governance is the competition among municipalities for business property because of the belief that business makes a net contribution to the local treasury. The public costs businesses impose are thought to be less than the taxes they contribute. Businesses, on the other hand, believe that they do not receive public service benefits equal to tax costs and therefore often oppose local public service expansion or locate in areas that provide service-cost packages more to their liking. Therefore, whatever the truth of the matter, incentives to businesses often play a significant restraining role on local financing. Independent of these competitive aspects, businesses are likely to cluster because of agglomeration economies. Consequently business will be distributed very unequally among municipalities, giving rise to severe differences in fiscal capacities among local governments.

The property tax on businesses is not well designed for the municipality, and it would be wise to move it to the metro or state level and then return the proceeds to municipalities and school districts. The tax yields could go into two funds; one for municipalities, to be distributed in proportion to employed population, and one for school districts, to be distributed in proportion to population or school attendance. Part of this reform is politically feasible now. There is strong support to remove business property from the school district assessment, put the property into a county, metro, or state pool for taxation, and then distribute tax revenues among school districts. It is less likely that a similar pooling could be arranged for municipal levies, primarily because of the ignorance about the relationship between the municipal tax payments of business and the municipal costs they generate. Though it is widely assumed that business is fiscally profitable, the research in support of this assumption is not very persuasive, and more study is called for before optimal programs to tax businesses for municipal purposes can be devised.

The property tax on residential housing has been more extensively studied, repeatedly criticized, and constantly eroded, but it still remains the mainstay of local finance. The major reform of this tax, proposed but not seriously acted upon in the United States, is a shift from a tax upon improvements to site taxation. Such a move would mean a relatively greater gain for the center city, where the land value per capita is greatest. The principal argument in defense of the shift to land taxation is that, with the tax on improvements removed, there would be an investment in the improvement of housing. The increase in investment per acre would mean that less land would be held vacant for speculative purposes and urban sprawl would be reduced—but so, of course, would the open spaces.

Site-value taxation has been accepted by economists for many decades as a tax on a surplus that does not distort the efficient allocation of resources. Despite this uniform approval by the profession, it has not received any support from men of affairs. Unfortunately site-value taxation became linked with single-tax reformism and assigned to the dustbin of crank proposals. Recently there has been an attempt to separate the economic sense from the evangelism, and we are probably at the point where a serious campaign by academicians and political leaders might engender significant movements to the transformation of the property tax to a land value tax. Research on the operations of the tax could be useful, but more important would be a demonstration.

The persistence of the property tax (land and improvements or just land) is partially a testimonial to the sluggishness of institutional change, but it also suggests that the tax has great virtues at the local level. An individual must make a locational decision among a set of governments. Municipalities can be distinguished by many attributes, prominent among them the package of public services and taxes, primarily property tax rate, they offer. The price an individual pays to reside in a municipality is the price of the land he purchases. The greater the difference between the benefits he receives from the public services and the costs he must incur to pay for the services, the more he will pay for the land. The taxes are assessed against the value of the property, and the taxpayer surplus, if one exists, becomes capitalized into the value of the property. The link between the property tax and the residential benefits-costs calculus provides a rough confirmatory test of the wisdom of the behavior of local officials. (Of course, the gains in value refer to land, not to property. Fiscal zoning refers to property, not to land.)

The tax competition we have stressed as a source of metropolitan fiscal problems has a positive aspect, too. Increases in land values are an incentive, and an "optimally organized" public sector would lead to increases in land values. If this were the full story, we could strongly urge that the property tax should be the mainstay of the system of local governments. Unfortunately, the gains in land-value increases can also be realized by beggar-thy-neighbor policies, which are destructive for all. Therefore a more complicated solution for metro fiscal policies is called for.

Income and Sales Taxes

Neither an income nor a sales tax is very satisfactory for any but the very largest cities. It is too easy to avoid such taxes by crossing municipal borders. These levies should be part of a state system of taxation and the basis of expanded state support of local services through intergovernmental grants.

The introduction, dispersal, and extension of sales and income taxes by states has been a drawn-out affair. In some cases, these taxes have been integrated with local levies on the same base (e.g., a local override on the state tax); in other instances they have been independently developed or there is a local tax on these bases without any state levy. It is common for the sales tax to be limited by the exclusion of commodities like food and clothing and for the income tax to be proportionate. Returns from both sets of taxes could be increased, together with an increase in progressivity in the entire structure, by the use of a sales tax credit against the income tax.

An extension of the sales tax to all commodities would greatly simplify compliance costs and would increase revenues. Of course, this transforms the tax, which is often proportionate because of these exclusions, into a highly regressive tax. The resistance to this extension is understandable, and therefore it would be desirable to link it with a sales tax credit on the income tax. Allowing a full tax credit against income tax would make the tax a progressive one. In fact, it would go beyond simple progressivity by allowing payments to be made to those who had negative income tax liabilities—in the case of welfare populations, for example.

Combining the credit sales tax with income tax would be of special value to the central city if state legislation permitted local overrides on the sales tax. Since the central city is overrepresented with low-income populations, a regressive tax would result in a greater revenue yield per capita; at the same time, the negative income tax feature would result in a cash transfer to the poor residents of the central city. The sales tax credit could be made more progressive by decreasing the amount as income increases. This is done by some states.

Conclusion

The jurisdictional fragmentation of the metropolitan area has been a major factor in the fiscal difficulties of cities. In turn, potential fiscal advantages to local governments have inhibited governmental reorganizations that might have efficiency and equity gains. There are some fiscal reforms, like the greater reliance on user taxes or the shift from property to land taxes, that should be advocated independent of the form of metropolitan government, but more important would be reforms consistent with a more effective government structure. Taxable resources are mobile, and so are the populations and land uses that generate demand for public services. A poorly designed system of taxing and service governments, like the ones typical of our metropolitan areas, will exacerbate the urban social and economic problems. We do not know the specifications of the optimal structure of metropolitan government, but there is little doubt that we can improve on the existing state.

3 Environmental Imperatives and Metropolitan Governance: The Case of Boston

MELVIN R. LEVIN*

Introduction

Before 1965 the physical environment was rarely identified as a critical social problem in urban texts, journals of urban affairs, or, most certainly, in legislation affecting the cities. Poverty and race relations were the victims of similar neglect up until the 1960s. By the end of the decade, however, past indifference in all three areas had given way to a flood of speeches, conferences, research studies, and additions to college curricula, but, lamentably, not much effective, substantive action.

This paper explores the feasibility of a metropolitan approach to environmental planning and program implementation. The discussion is based on four propositions. First, although there is a legitimate role for metropolitan action aimed at alleviating environmental problems, there is no special or unique reason why such a role must necessarily be related to any one type of metropolitan governing body. Metropolitan-wide governments have played more or less active roles in several countries in formulating environmental programs. Stockholm, Copenhagen, Paris, Amsterdam, and London have been much more careful than most American cities to preserve historic buildings, to maintain, in good condition, substantial inner-city parks and parkways, rivers and canals, and to set aside and protect green areas on the outer fringe.

*Chairman, Department of Urban Planning and Policy Development, Rutgers University.

These efforts reflect in large measure the presence of different varieties of strong central governmental agencies with effective power over land use in the entire metropolitan area. In contrast, the American problem is one of prerequisites—that is, the granting of enough public power on a scale wide enough to control development. Other Western nations take such power for granted.

Second, it is assumed that unless there is far more fundamental change in the U.S. federal system than now seems in the cards, the lion's share of the funds, of the initiative, and of the ground rules for environmental programs will emanate from the federal government. A cursory look at the present program activity of various government levels quickly shows the preponderance of the federal role, at least in those areas in which any government action has been undertaken (see table 1). This is most obvious in the setting of standards, providing or initiating funding, and program development and sponsoring research. The federal government is also able to take an active part in such areas as population control, which lesser jurisdictions have not yet accepted as a legitimate government function.

Third, the very newness of the widespread, serious concern for the environment and the large number of significant "unknowns" militate in favor of a certain looseness in organizational design and structure. The returns are not likely to be in for some time as to optimum governmental approaches to problem solving.

Finally, the great size and complexity of the nation suggest that different forms of organization may be appropriate in different areas. In addition, the optimum areal unit may vary in size, depending on the type of environmental issue involved. For example, Maryland's Environmental Services Act of 1970 provided for immediate establishment of service regions covering the entire state but recognized that regions may differ in size, depending on function. Maryland's wastewater purification regions and solid waste disposal regions need not be identical.[1]

Seven environmental factors—population, water pollution, land use, air pollution, noise pollution, solid wastes, and esthetics—have been selected for consideration. For the purpose of field testing, the focus is on metropolitan Boston, which has the twin advantages of a long history in certain types of metropolitan governance and proximity to the author (who was on the faculty of a Boston area university at the time of the major research). Each of the seven environmental areas will be discussed briefly in terms of the nature and severity of the problem, of the costs and benefits involved in remedial programs, and of the experience and prospects in each area for metropolitan Boston.

A summary of the principal general findings and conclusions is presented in table 1. It will be noted that some metropolitan agencies currently play a

[1] Environmental Services Act of 1970 (Maryland), July 2, 1969, p. 8.

minor planning role in two of the seven areas—land use and solid waste—while program implementation by metropolitan agencies has been largely confined to land use. However, two caveats are in order. First, a number of the powerful metropolitan agencies operate outside the environmental tent but nevertheless have a great impact on environmental patterns and problems. These include metropolitan transportation, parks, and public utilities agencies. Second, care must be taken lest metropolitan *approaches* be used as synonyms for metropolitan *government*. In actuality, the governance of metropolitan areas takes a number of forms, from refusal to admit the existence of a distinct metropolitan entity beyond the thin cosmetic facade needed to qualify for federal funds, to full-fledged metropolitan government. Most larger metropolitan areas, however, like Boston, present a complex governmental pattern that has evolved piecemeal over the years. On the assumption that some type of organic evolution is more probable than a sweeping reshaping of governmental affairs in metropolitan areas, the Boston example is more widely relevant than such governmental models as Toronto, Miami, or Nashville.

Principal Metropolitan Agencies

In practice, the principal agencies or "actors" in the Boston metropolitan area continue to function in their traditional, fragmented manner, despite membership in the new state "umbrella" agency, established in 1969.

Metropolitan Area Planning Council (MAPC)

Established by act of the legislature in 1963 with strong and continued sponsorship by the federal Department of Housing and Development, the MAPC includes 100 member communities in metropolitan Boston with a combined 1960 population of over 3 million and is authorized to assess member communities 5 cents per capita. Under the governance of its member communities (with Boston having a strong role), MAPC decided in 1969 to withdraw from direct state control. It can now deal directly with federal agencies for contract studies and other inputs.

The MAPC has sponsored a number of research studies in population, economic development, housing, employment and manpower, health and education facilities, solid waste disposal, open space and recreation, regional sewer and water facilities, aid to local government, and transportation. The MAPC's principal initiative actions have been in the fields of land use (i.e., open space), sponsorship of a proposed regional airport, and a regional incinerator program for disposing of solid wastes. Much of the staff's time is spent reviewing development plans of member communities and providing technical assistance when the communities request it. Overall, the agency

TABLE 1. Intergovernmental Relations and Environmental Programs

	Status	Research and planning	Program operation and implementation	Program coordination and standards	Source of funding	Major obstacles
Population control	U.S. zero growth likely by 1990–2000; growth rates highest among low-income groups, but tending to decline	Primarily foundation, academic; increasing federal	Few government programs; some family planning aid, etc.; more private programs; key is private-personal choice; public & media choices helpful	Little in evidence; some federal (poverty programs)	Primarily private; increasing federal	Inadequate public awareness and knowledge, particularly among low-income families
Water pollution	Serious problems nationwide. Few major improvements; some areas deteriorating	Primarily federal, much regional (e.g., river basins)	Federal, partly by prodding lagging states; some regional (river basins, interstate compacts)	Federal; river basin commissions important	Federal & state; some local	Shortage of funds to cope with massive backlog; stagnant technology
Air pollution	Serious cause for concern in most large urban areas; many isolated local areas affected; some improvements noted	Federal	Federal, partly through regional airsheds & state; minor local efforts in large cities	Federal	Federal; some state	Rapid growth of emission sources; weak federal leadership in research & funding; no consensus on standards

Noise pollution	Serious in localized areas; generally worsening problem	Limited federal	Some federal, state & local; much attention centered on airport areas & industrial noise	Some federal (airports & industrial noise)	Minor federal	Increase in amount and variety of emission sources; weak federal leadership in research funding; no accepted technical standards
Solid waste	Serious problem nationwide; rapidly worsening in medium-to-large urban areas	Federal, state; increasing municipal	Municipal; some subregional; incipient state	Some state, some subregional	Primarily municipal; some federal	U.S. affluence, rapid rise in volume; rising labor costs; more feasible to produce new products than to reclaim & recycle; major advances in technology needed & possibly manufacturing product guidelines
Land use	Serious problems in many areas; some local improvement	Federal, state, metropolitan, municipal, private	Municipal; some metropolitan. Growing federal & state role in setting standards & guidelines	Primarily municipal; some state & metropolitan	Federal, state, municipal, some metropolitan	Fragmented responsibilities; intermunicipal rivalry; no consensus on criteria & standards
Esthetics	Cause for concern in many areas, exacerbated by comparisons with best foreign & U.S. examples	Some federal; some foundations, professional organizations	Mostly local; some federal	Mostly local; some federal (e.g., interstate highways)	Some federal, minor city, state	Lack of public awareness; no consensus on criteria & standards

seems to have only a limited impact on metropolitan development patterns, an assessment that may well apply to most metropolitan planning agencies.

Metropolitan District Commission (MDC)

Massachusetts pioneered in the regionalization of public services in metropolitan Boston with the establishment in 1889 of the Metropolitan Sewerage Commission. In 1893 a Metropolitan Parks Commission was created, followed in 1895 by a new Metropolitan Water Commission. In 1919 the three commissions were consolidated into the Metropolitan District Commission, a state agency whose commissioners are appointed by the governor and whose budget is controlled by the legislature in the same manner as those of other state departments. A total of 52 communities belong to one or more of the MDC districts, but community membership in the MDC varies: 32 communities are in the water district, 41 in sewerage, and 37 in parks. The MDC builds trunk sewers and pumping stations, plans and operates reservoirs, aqueducts, and mains, and has established a system of parks, green-belt parkways, and a number of recreation areas.

Transportation Agencies

In his New York study, Robert Wood labeled the Port of New York Authority and some of its sister regional enterprises, mostly in transportation, "the metropolitan giants."[2] Metropolitan Boston is fully stocked with similar behemoths whose transportation activities have a direct bearing on land use and other aspects of the environment. Four of these agencies have specifically metropolitan responsibilities.

Massachusetts Bay Transportation Authority (MBTA). Created in 1964 by act of the legislature, the MBTA incorporated and expanded the original metropolitan transportation system (created in 1947) known as the Metropolitan Transit Authority (MTA). The act provides that the authority may coordinate services of existing public and private transportation, lease or purchase other transportation facilities within the area, and receive additional funds through an increased cigarette tax and a short-term subsidy. The MBTA is governed by a five-member board appointed by the governor and an advisory board on which each of the member cities and towns sits and has voting power commensurate with its financial contribution.

Massachusetts Port Authority. Created in 1956 by act of the legislature, the Port Authority provides facilities and services to promote trade and commerce for the economic benefit of the citizens of the commonwealth. The

---

[2] See Robert C. Wood, "The World of the Metropolitan Giants," *1400 Governments* (Cambridge: Harvard University Press, 1961), chap. 4, pp. 114–172.

Port Authority, although within the Department of Public Works, is not subject to its control and is financed exclusively through bond issues and user charges. The authority is responsible for a major bridge leading to the north of Boston and for Logan International and Hanscom airports, as well as for the Port of Boston. The agency has concerned itself mainly with the aviation facilities, which are generally thriving, and is involved in real estate ventures, including promoting construction of a trade center in the fashion of its sister agency in New York. Despite its name, it has had little success stimulating growth in Boston's harbor or in reviving its fishing industry.

Massachusetts Turnpike Authority. Created in 1952 by act of the legislature, this agency is also within the Department of Public Works but operates independently of its jurisdiction. It is responsible for operating two tunnels linking the city of Boston to the North Shore, in addition to the east-west turnpike between downtown Boston and the New York state line. The agency consists of three members appointed by the governor and is financed by revenue bonds and revenue from the turnpike.

State Department of Public Works (DPW). Although it is a state agency like the other transportation units, the DPW is not similarly structured: in addition to having responsibility for statewide transportation planning and construction, it has none of the trappings of an authority, including the (often minimal) element of community or public representation at the governing level. Nevertheless, in practice the DPW operates in much the same quasi-autonomous fashion, due largely to its enormous financial resources, including construction and maintenance contracts, which give it a powerful voice in the legislature. In theory, the DPW, like many other state agencies, has regionalized its activities. Massachusetts is a small state, however, and, in practice, since all of the headquarters offices are in Boston, all of the agencies have centralized decision making, often down to quite minor details.

Environmental Agencies

State responsibility for environmental programs has been scattered among several agencies. In addition to the MDC, the MAPC, and the transportation agencies (MBTA, Massachusetts Port and Turnpike authorities, and the DPW) already mentioned, a number of other agencies have been important in one or more areas of the environment: the Department of Public Health, the Department of Natural Resources, the Division of Water Pollution Control and the Division of Water Resources of the Water Resources Commission, the Registry of Motor Vehicles, and the Outdoor Advertising Board. The DPW, in addition to its transportation interests, was given statewide responsibility for solid waste disposal. Some of these agencies have legal authorization for their work; others have taken an interest in the field and have attempted to carve out for themselves some area of responsibility. Still other agencies have been forced into the environmental field because their operations have produced one type of pollution or another.

Departments of Public Health and Natural Resources. Most of the state power over pollution control programs is vested in these two agencies. Both departments have statutory obligations to prosecute violators of state environmental laws. Of the two, Public Health's role is more comprehensive, with responsibility in the areas of water, air, and noise. The department's Division of Environmental Health supervises the testing of public water supplies. The department is responsible for implementing the air pollution control program of the thirty-community Metropolitan Air Pollution Control District. Creation of the Metropolitan Air Pollution Control District was required by federal law and is two-thirds funded by the U.S. Department of Health, Education, and Welfare.[3] Created by the legislature, the district is for all practical purposes part of the Public Health Department. Virtually coterminous with the MDC's Parks District, the air pollution operation receives one-third of its funds from the legislature with the state treasurer levying annual assessments on the member communities, after the practice of the MDC and the MBTA. New, small, and understaffed, the agency has been of limited effectiveness. At the same time, Public Health is establishing air pollution control districts in other areas of the state.

The Department of Natural Resources (DNR) is most active in the area of land use. This is not, however, a comprehensive responsibility, but a limited one relating in particular to the state's public lands, water resources (the Great Ponds and rivers), and the inland and coastal wetlands. Development of privately owned dry land is largely beyond the department's control or influence. The state's wetlands legislation is innovative and has had significant success in protecting certain areas considered by the state to be unsuitable for development. In addition, with the advice and financial support of the Conservation Commission legislation, the department has helped many municipalities purchase or protect lands for environmental reasons. The department's other area of major interest is outdoor recreation. DNR now operates a large number of recreational facilities, from ski slopes to camping sites, throughout the state.

State Division of Water Pollution Control. Although located within the Department of Natural Resources, the Division of Water Pollution Control and the Division of Water Resources are responsible only to the Water Resources Commission. The division has wide, but only partially used, powers in broad areas affecting rivers, lakes, and harbors. A more limited role is

---

[3] Air Quality Act of 1967, P.L. 90-148, November 21, 1967.

played by the MDC, which polices pollution in the Charles River and the Charles River Basin in addition to water bodies that are part of the MDC water district, including about 50,000 acres of land and water extending as far west as the Connecticut River.

Registry of Motor Vehicles. Empowered to regulate automobile exhaust systems to prevent excessive smoke or noise, the registry in the past operated independently of both Public Works and Public Health. But the recognition of the major role played by vehicle exhausts in air pollution led, in the early 1970s, to attempts to coordinate the registry's activities with those of the Department of Public Health. In practice, however, the key role in dealing with auto emissions has been that of the federal Environmental Protection Agency.

Within the DPW is the Division of Solid Waste Disposal. The division, not fully organized or funded in mid-1971, devoted much of its attention in 1972 to a research study focused on the scope of the solid waste problem and the lack of regional responses in the metropolitan area.[4]

The cooption of solid waste by the state highway agency reflects a belated but nevertheless successful attempt by the DPW to become the leading agency in this field. The DPW's solid waste division was in fact a hasty response to an MDC proposal to provide a system of incinerators for the eastern Massachusetts region. The DPW proposed to build and operate a series of regional sanitary landfills. Although discussion often centered around the merits of landfill versus incinerators, the issue of state versus metropolitan responsibility was clearly one of the deciding factors in a controversy between the two politically powerful agencies.

Outdoor Advertising Board. The state's direct role in esthetics is limited, but it does have major leverage over the billboard industry. In practice, however, the Outdoor Advertising Board, created in 1955 for "regulating and controlling, in the public interest, the erection and maintenance of billboards, signs or other advertising devices," has been better known for its deep sympathy with and understanding for the outdoor advertising industry than for its stern regulation of roadside eyesores.

This brief outline provides a basis for examining existing and potential roles for metropolitanism in environmental planning. One final point should be made clear: the Metropolitan Area Planning Council is the only substantial metropolitan agency that makes more than a pretense of control and direction by area representatives. Every other actor on the metropolitan scene is, in effect, a branch of state government, whatever the ostensible administra-

---

[4]*Raytheon Service Company Report*, prepared for the Bureau of Solid Waste Disposal, Massachusetts Department of Public Works, May 1972.

tive regionalization for metropolitan Boston. This fact has many implications, by no means limited to the Boston area.

State Agency Reorganization

In a major effort to put an end to this hodgepodge of program responsibility, the legislature enacted in 1969 a bill to reorganize Massachusetts' more than 200 executive departments and agencies into 10 secretariats, effective April 30, 1971. In theory, this provided for more effective environmental action, since it focused responsibility and control for environmentally related programs in the new Executive Office of Environmental Affairs. This super-agency has as its mission

> to protect and improve the quality of our natural environment and the resources which together constitute it, and to improve the public's opportunity to enjoy and exist healthily in the environment, by controlling the man-made despoilation of our resources and directing growth and development along planned lines which will preserve for all time an ecologically sound and esthetically pleasing balance of naturally-occurring resources.[5]

In order to accomplish this objective, the secretariat brought together under one roof the following agencies: Department of Natural Resources, Outdoor Advertising Board, Water Resources Commission, and part or all of the Metropolitan District Commission. The MAPC, however, was placed under the Executive Office of Communities and Development; the Registry of Motor Vehicles under the Executive Office for Public Safety; the Department of Public Works under the Executive Office of Transportation and Construction; and the Department of Public Health in the Executive Office of Human Services. Obviously, considerable fragmentation of responsibility continues to exist. Although this reorganization shows promise of improving cooperation among the many agencies having responsibility for metropolitan areas, this promise has yet to be realized, partly because the reorganization has not been fully implemented.

The Metropolitan Environment and the Organization of Government

Population

Of the factors bearing on the environment, population is probably the most crucial, since it affects every aspect of environmental pollution. Popula-

---

[5] Massachusetts, Executive Office for Administration and Finance, *Modernization of the Government of the Commonwealth of Massachusetts as Enacted 1964* (August 1969), p. 11.

tion control involves two key issues: (1) influencing the overall size of the population and (2) effecting population distribution consistent with environmental objectives. The United States seems well on its way toward achieving the first objective. By the early 1970s, birthrates had dropped below depression levels, and zero population growth seems a strong possibility before the end of the present century. Nevertheless, the impact of the population of the United States on its natural resources is tremendous—its 6 percent of the world's population currently uses more than 40 percent of the world's scarce, nonreplaceable resource output. Hence the question of population control is as relevant a concern in the United States as it is in less-developed countries.[6] Indeed, at best we can still expect a last great surge in population growth, another 50 million people by the year 2000.[7]

Americans are becoming sensitive to the possible dimensions of population growth and its consequent impact on the nation's resources. Some experts warn that the United States has already passed its optimum level, while others feel the point has yet to be reached. Zero growth rate is possible in the next two generations, but without substantial reductions in generation of pollutants, even modest population growth can pose a serious threat to natural resources and environmental quality.

Whatever the accuracy of population forecasts—these have proved notoriously inaccurate in the past—changes in birthrates, the chief determinant in our population growth, are based on a multitude of private decisions. Government's role can be significant in providing family allowances, low-cost abortions, adequate housing, day-care centers, and family planning clinics. With modern contraceptive techniques, however, children should be a matter of choice rather than inevitable necessity. In the 1960s, more families began to choose greater affluence, based in part on caring for only two, or at most three, youngsters, over a lower living standard for four or more. Thus, one major pollution-oriented problem in the United States seems on its way to mitigation, largely as a matter of free, family choice.

Among low-income families, however, large numbers of children may lock the family into poverty and, through parental and governmental neglect, a large proportion of the poor may thus be condemned to existence in a generational poverty cycle. We still have not made adequate arrangements for chil-

[6]*Environmental Quality*, First Annual Report of the Council on Environmental Quality (Washington, D.C.: Government Printing Office, 1970), p. 14.

[7]*Population and the American Future*, Report of the Commission on Population Growth and the American Future (New York: New American Library, 1972), pp. 9–21. On the basis of a norm of two-child families and a continuation of immigration at the 400,000-per-year level (excluding illegal immigrants), the U.S. population would grow to 271 million by the year 2000, as compared with 220 million in 1973. Three-child families plus continued legal immigration would raise the U.S. total to 322 million.

dren of the poor to make possible a higher standard of living for this group and thus break the cycle of poverty, which is closely linked with large, often unwanted families.[8] Part of the answer lies in providing females in low-income families with free access to the birth control information currently available to middle- and upper-income women. This is the goal of the Family Planning Services and Population Research Act supported by the Nixon administration, which established an Office of Population Affairs to give low-income families priority in receiving voluntary, cost-free family planning services. During 1971-1973, $26 million were authorized for this program, but none of this can be spent on programs where abortion is a method of family planning.[9]

This program could result in better living conditions for smaller families, and reduced rates of dependency, welfare, and crime. A well-fitted pessary or low-cost abortion can easily save thousands of dollars in taxes, in addition to increasing the chances for children in two- or three-child families to climb out of the grip of poverty.

A more difficult population problem is the tendency of population growth to concentrate in a relative handful of megalopolitan areas, particularly on the coasts, leaving most of the nation stable, declining, or empty. To the degree that overcrowding is a worrisome phenomenon, some of the clouds seem to be lifting. Some convergence in population densities has been taking place in many areas; central-city densities are heading downward to the 10,000-15,000 per square mile range, and suburban densities reaching toward the 2,000-5,000 level. The population in suburban communities surpassed central-city populations during the decade of the sixties.

One way to cope with the density problem is by setting population ceilings for portions of the metropolitan area. The relatively slow growth of metropolitan Boston and its large amount of open territory remove the urgency of imposing population limits of the kind proposed for Los Angeles and Denver.[10] There is a case, however, for allocating population growth within the metropolitan area: some communities that have engaged in restrictive zoning practices may need to increase their population densities to take the pressure off communities that seem to be having serious growth troubles. Although this objective cannot be pursued directly through an area or state

---

[8] Over half the children born to black mothers with less than any high school education are "unwanted," compared with 7 percent among women with four or more years of college (*Population and the American Future*, p. 164).

[9] "Family Planning," *Congressional Quarterly Weekly Report* 27, no. 48 (November 20, 1970): 2817.

[10] See Steven V. Roberts, "Some Areas Seek to Halt Growth," *New York Times*, March 14, 1971, p. 1.

population policy, it may be approached by indirection: the legislative success in striking down Massachusetts' suburban snob zoning may go far to achieve a partial redistribution of area population.

The chief population issue in state government has been reforming abortion laws and removing restrictions on dissemination of contraceptive information and materials. Massachusetts has loosened its restrictions in the latter area, but in practice administrative procedures and facilities are somewhat more limited and restrictive than in New York and other states. There is no large-scale abortion counseling–entrepreneurial industry on the New York model, for example. Municipalities, on the whole, have been supportive of federally aided family planning efforts through poverty programs. Some of the lack of resistance may be due to the fact that no local funds are involved and to resentment about payments to welfare mothers, who allegedly enlarge their families in an irresponsible manner in the knowledge that welfare will foot the bill. Some feel that there may also be overtones of racism, since family planning may decelerate population growth among the blacks and the Puerto Rican population.

Where does this leave metropolitanism? It would not be wholly accurate to suggest that there was no potential role at all, especially if control of area population size and growth rates becomes an accepted goal. A number of foreign nations have taken steps to limit the size of their larger urban agglomerations. Should population objectives become part of a national urban policy or of metropolitan goals, land-use policies could be directed toward channeling population growth in much the same manner as wealthy suburban communities have restricted an unwanted influx: zoning and the development of public facilities and services can be instruments in accelerating or decelerating housing construction and economic expansion.

However, major interests in U.S. metropolitan areas view rapid population growth as synonymous with economic health and social vitality. The development proposals advanced by the MAPC in the 1960s are little more than attempts to accommodate a rate of population growth over which the metropolitan area is assumed to have no control. Indeed, the MAPC has billed its population studies as forecasts in which the chief variables relate to birthrates and migration, both viewed as natural, uncontrollable phenomena.[11] Realistically, perhaps metropolitan governmental responsibility should focus only on the area of population distribution, leaving problems of total growth to be solved by national population, economic, and migration policies and private, family choice.

[11] See Metropolitan Area Planning Council, *Economic Base and Population Study*, vol. 3, *Population Projections for the Eastern Massachusetts Region* (1966).

Water Pollution and Water Resources

Water pollution has probably received more attention than any other aspect of the environment. Solutions to the water pollution problem are expensive, and tangible returns may appear to be small: clarifying turbid water, removing odors, and restoring fishing and swimming are limited benefits. The principal economic argument for water pollution abatement rests on a longer-term basis—the menace to the balance of nature posed by continued water pollution threatens the continuation of the species.

Until recently, Americans acted as though their water resources had an unlimited capacity to absorb wastes. Dirty water, like dirty air, was considered a sign of progress and economic prosperity. No area of the country has escaped the problem in one form or another. The hardest hit have been the Northeast and the Great Lakes region, largely because of extensive urban and industrial development during the past fifty years. Other areas suffer from specific sources of pollution, most notably coal mining, an industry that has ravaged the hills and poisoned the streams of Appalachia.

The principal industries causing serious pollution include paper, organic chemicals, petroleum, and steel. Most industrial wastes can be handled with present technology, but some types of industrial pollution present difficult abatement problems and new products (particularly nonbiodegradable plastics) continually being developed present new challenges.

Only 45 percent of the wastes treated by municipal treatment systems comes from homes and commercial sources. Of the total population, one-third is served by a system of sewers and an adequate treatment plant, one-third is not served by any treatment system at all, about 5 percent is served by sewers without any treatment, and the remaining 32 percent have sewers and an inadequate treatment system. This unsatisfactory condition is likely to continue, unless drastic efforts are made to cope with present demand, because waste loads are expected to quadruple in the next half-century.

Agricultural pollutants include animal wastes and the discharge of fertilizers and pesticides. Chemical fertilizers, used increasingly in recent years, have caused severe water pollution problems in some areas.

Another source of water pollution that has surfaced in recent years is the oil spill. With the grounding of the *Torrey Canyon* in 1967, the Santa Barbara offshore oil leak in 1969, and the oil spills in the San Francisco Bay region and off the shores of New Haven in 1970, oil pollution has suddenly become a serious national and worldwide problem.

The effects of these pollutants are felt on human health, on the availability and quality of outdoor recreation, on commercial fishing, on agriculture, and on industrial and municipal water supplies. Attempts to put a price tag on the costs of clean water estimate that between 1970 and 1975 $10 billion will be needed for new municipal waste treatment plants to meet water quality standards—about $2 billion a year. In addition, operating costs are estimated to rise from $410 million a year in 1969 to $710 million in 1974. And the problem of separating sewer and storm overflows has been estimated to cost between $15 and $48 billion.[12] These can be considered only very rough estimates at best. Depending on the solution chosen, widely differing costs will be encountered. Nevertheless, water treatment will be extremely costly, especially in terms of initial capital investment. This cost, which must be borne by government, industry, and the individual citizen, can be justified, however, if only because without it life may become impossible. But even these large sums of money will not produce a nation of clean rivers. The estimates are based on work that will bring water quality up to artificial standards already attacked as inadequate. As a result, cost estimates for cleaning up the nation's rivers and harbors vary by a factor of nearly eight. In any event, the total costs will be very large; in late 1973 the federal Council on Environmental Quality estimated the necessary outlay to meet the criteria set forth in the Water Pollution Control Act of 1972 at $121.3 billion during the period 1972-1981.[13]

The federal government did not act to control water pollution until 1948, when the Federal Water Pollution Control Act was passed. This act authorized planning, technical assistance, grants for state programs, and construction grants for municipal waste treatment plants. Subsequent legislation in 1965 created the Federal Water Pollution Control Administration and called for the establishment of water quality standards and implementation plans for clearing up all interstate and coastal waters.

In Massachusetts, responsibility for ensuring adequate water resources of a high quality is spread among a number of agencies under the general aegis of the Executive Office of Environmental Affairs: the Department of Natural Resources, the Water Resources Commission (WRC), the Department of Public Health, the Metropolitan District Commission, and the Department of Public Works. In addition, the MAPC has done a study of regional sewer and water facilities.

The DNR, WRC, and DPH are the major state agencies functioning in this area. DPH is primarily concerned with the health aspects of water quality, testing, and standard setting. DNR and DPW have jurisdiction over various waterways, including Boston's inner harbor and the rivers that drain into it, and beaches and harbors. The DNR and WRC set water quality standards and acquire and restrict the use of coastal and inland wetlands.

Along with the U.S. Army Corps of Engineers, the MDC has specifically metropolitan responsibilities over Boston's inner harbor; it can restrict and

[12]*Environmental Quality*, pp. 42–43.
[13]Gladwin Hill, "Estimate of Pollution Control Costs Pared," *New York Times*, September 18, 1973, p. 26.

acquire marshes and banks and make channel improvements. The sewerage and water systems of the MDC are probably the most successful examples of metropolitan governance in the Boston area, although the MDC is not a genuine self-governing metropolitan agency. Few urban areas can boast a water resource system as vast as that which runs from the Quabbin Reservoir in central Massachusetts to Boston and neighboring coastal communities.

With the exception of the MDC, a comprehensive listing of Massachusetts environmental legislation failed to show any substantive role for metropolitan action. Moreover, the emphasis in the new legislation enacted in the late 1960s and early 1970s was on strengthening the powers of the state in this field, as in others.

In the late 1960s and early 1970s, DNR (subsequently made part of the new Executive Office of Environmental Affairs) engaged in a sporadic conflict with the City of Boston over the future of key harbor islands. Specifically, the city's interest was in the direction of urbanization for a new community or other purposes, while DNR stressed the need for recreational preservation with only modest development.

The tougher environmental regulations enacted in 1969 and 1970 included posted bonds of at least $25,000 for tankers discharging oil cargo, rules governing sewage treatment prior to discharge into commonwealth waters, and regulation of marina and boat sanitation facilities.[14] The legislation also included a new $150 million state construction grant program and $1 million a year for research and development for better waste treatment methods.[15] The primary responsibility is vested in the Division of Water Pollution Control. Other significant legislation includes a $250 million bond issue for construction grants and a $25 million low-interest loan program for industry.

Water pollution has generated the most attention in terms of both heightened public awareness and its needs for massive capital investments. The cost of cleaning up rivers, lakes, and harbors in older regions like New England may eventually overshadow the vast costs of the interstate highway program. The agency which preempts planning, research, problem solving, and contract awarding in this field may wield an influence analogous to the agencies ruling the highway empires. At present, this turf has been staked out almost everywhere by the traditional water resources agencies, along with such comparative newcomers as the New England River Basins Commission, the New England Interstate Pollution Control Commission, and the trouble-ridden New England Regional Commission. The key to power—control over

---

[14] Massachusetts, Office of Comprehensive Health Planning, *Compendium of Environmental Legislation* (with addendum: Massachusetts Acts of 1970), June 1970.

[15] Thomas C. McMahon, "The Water Pollution Control Program in Massachusetts," *Community Planner* (Massachusetts Department of Community Affairs) 3, no. 1 (January 1971).

contract awards—remains in the hands of state agencies and their two partners, the federal agencies, which provide desperately needed funds, and the hard-pressed local governments (and reluctant private corporations), which are being prodded to build waste treatment facilities.

Where does this leave metropolitan agencies? At the moment, the Boston region's planning agency (MAPC) is a non-starter. Its role in water pollution control is limited to passing comments in its reports and conclaves urging priority action on this front.[16] It would not be difficult to identify a meaningful role for a legitimate metropolitan agency in this field in research and development, planning, and, indeed, in contributing to program implementation. Instead, MAPC's role has been restricted largely to the review and advisory function required as part of the federal checkpoint procedure, although its sounding of the alarm bell has furthered the expansion programs of operating agencies. Under review responsiblities, metropolitan and regional planning agencies are required to certify that proposed municipal facilities are consistent with regional plans before federal grants and loans can be awarded for such facilities as sewage treatment plants. Practically speaking, water pollution control has been left to the federal and state line agencies, including that long-established example of functional metropolitanism, the MDC (with federally sponsored, interstate river basin commissions playing a modest role) and to local public and private consumers and polluters.

Land Use

The chaotic urbanization of our time and place is partly associated with the expansion and distribution of the population, although enormous problems can develop even where there are very few people. Some industries are major land polluters: not only coal mining in West Virginia and Pennsylvania, but also countless junkyards, chemical plants, fertilizer factories, and pulp mills blight the landscape—and often the air and water—over large areas.

The trend is not all downhill. Some tendency toward efficient and attractive use of land is observable in a number of industrial parks and shopping centers, as well as in residential developments. Nevertheless, many of the mistakes of the past can be rectified only at enormous cost. The price tag on cleaning up slum housing alone is likely to exceed a trillion dollars, much of

---

[16] See, for example, *Guides for Progress: Development Opportunities for Metropolitan Boston*, prepared by Metropolitan Area Planning Council (Eastern Massachusetts Regional Planning Project), April 1968, chap. 3, p. 11. The MAPC issued a report identifying the gap between regional water supplies and projected demand, recommending a substantial expansion in the MDC service area (Greater Boston Chamber of Commerce, *"TASK FORCE": Boston Government/Regional Delivery Systems: A Metropolitan Boston Model for Regional Economic Development and Delivery of Services* [Boston: March 22, 1972], pp. 4–5).

which may have to be provided by government agencies. Other efforts to remove and regroup inefficient and often dangerous commercial strip development, substandard resort housing, deteriorating waterfront areas, and blight, which afflicts rural as well as urban areas, may run into the billions of dollars.

History suggests that motivation and life styles are as critical in cleaning up slum areas as changes in the physical setting. Specifically, this fact poses two alternatives.

The first is to adopt a *suburban strategy*, concentrating programmatic efforts in the areas where most of the new urban development is taking place. The move to the suburbs continually devours open land for housing, shopping centers, highways, airports, and light industry, often sacrificing recreation and park needs. Such a policy would, in effect, write off the inner-city slums as beyond redemption at this time. The high cost of maintaining slums as they continue to rot away would be viewed as a "given," an inescapable component of modern budgeting, and certainly much less expensive than mounting and financing programs that flounder around in vain attempts to reconstruct blighted low-income areas. This line of reasoning is sometimes linked with corollary efforts aimed at dispersion, a kind of quota system under which suburban communities would be required to set aside some land for construction of housing for low-income families. A "new towns" policy could be an important part of such a strategy and could control the development of new, balanced suburban communities. European experience offers several examples worth citing, the most notable of which is Finland's Tapiola. A nonprofit group of welfare and labor organizations during the postwar housing shortage provided the initial support. Since then the project has been largely self-financing, although it has continued to receive subsidies available to any nonprofit organization providing low-income housing.

> Tapiola's high planning standards, varied housing designs, sound industrial base, and amenities greatly strengthened its appeal to both white and blue collar classes, and this permitted the Foundation to surmount fiscal and other difficulties.[17]

At least one democratic nation has demonstrated that it is possible to develop well-conceived, economically feasible, and architecturally attractive new towns that offer housing for a broad socioeconomic spectrum of the population.

Massive head-on intervention to *rebuild the slums*, quite possibly with greater emphasis on home and apartment ownership, characterizes an alternative approach, one that would have to be accompanied by large-scale suppor-

---

[17]Advisory Commission on Intergovernmental Relations, *Urban and Rural America: Policies for Future Growth* (Washington, D.C., 1968), p. 67.

tive services to ensure that profiteering, vandalism, neglect, litter, and other human problems did not create instant slums out of new buildings. The high costs of reconstructing slum areas could soak up most of the funds available for urban development, a fact that presents serious questions of priorities with major racial, class, and political implications, and the likelihood of confrontations.

Regardless of the direction chosen (and evidence seems to point to an uneasy compromise, with a continued "tilt" toward the suburbs), more active government efforts to control land-use patterns were in prospect in the early 1970s.

The control of land-use development presents an anomalous picture. While the primary responsibility for planning, zoning and subdivision control, and provision of key public facilities is vested in a multiplicity of local government agencies, major development determinants (e.g., highways) are the responsibility of state agencies. Though frequently required to consult with local governments before reaching final decisions affecting communities, in practice the transportation agencies and other major state and federal agencies often prepare and implement plans with little reference to local desires. The degree of resilience in agency plans may vary, depending on local political strength and other factors, but, on the whole, local governments formulate land-use plans within the framework of a number of "givens," not the least of which are the development programs formulated by key state agencies.

Local land-use policies are strongly influenced by private developers, although the relationship varies from community to community. Some of the more affluent suburbs can afford to be selective; their poorer neighbors may be highly responsive to all sorts of proposals advanced by developers.

State government has become increasingly active in the land-use area, either by raising environmental challenges to prospective developers, as in Vermont, or by attempting to formulate and implement comprehensive land use controls, as in Hawaii and California. By the end of 1973 the passage of federal legislation aimed at stimulating serious state land-use planning seemed a certainty.

Finally, the federal government operating through a variety of not-always-consistent programs also exercises powerful influences on land-use patterns. The rate of housing growth, which in turn is keyed to suburbanization, largely depends on such federal programs as housing interest rates and the federally aided highway program, while the future of core communities may rest on a variety of federal aid programs ranging from urban renewal to public transportation.

Out of an estimated million acres in the Boston metropolitan area, the MDC already controls 21,000 acres of water, parks, and recreation facilities, and the state Department of Natural Resources controls almost 12,000 acres of state forests and parks. Each of these agencies has a master plan of its own for acquisition and development. Further, all of the metropolitan area cities and towns, which own a total of 55,000 acres of open space, have prepared plans for land-use development. Meanwhile, as noted, the major transportation agencies have all prepared their own master plans.

Given this complex of strong forces already operating in the land-use arena, what remains for the metropolitan agency? MAPC has chosen to pursue a number of strategies.

(1) *Research.* MAPC has sought to become a primary source of definitive information on area needs and problems in the hope—so far unrealized—that agencies and communities would develop plans and programs on a consistent foundation.

(2) *Accepting "givens."* In the Controlled Dispersal Development Guide and Composite Development Guide presented in *Guides for Progress*,[18] MAPC incorporated virtually all elements of the master highway and public transportation plans, community housing trends, master plans, and the like prepared by other agencies as basic elements in its alternatives. In this sense, MAPC has staked out a role as a synthesizing rather than a determining factor in area development patterns.

(3) *Limited intervention.* MAPC has itself attempted to reshape the area's land-use development pattern to a limited extent by presenting its own proposals for adoption, including an open space and recreation study that called for planned acquisition and development on a larger scale than proposed in MDC, DNR, or other plans.[19] As noted, MAPC also financed an airport study that proposed construction of a second international airport in an affluent, thinly populated community to the southwest of Boston. The proposal, which called for the acquisition of 10,000 acres for the airport and would have had broad implications for land-use patterns in a much larger impact zone, was soundly defeated before it reached the status of a serious piece of legislation. A bill calling for further study of alternative sites for a second major airport was filed in the legislature in 1970.[20]

[18] *Guides for Progress*, chap. 4, pp. 35–40.

[19] Massachusetts, Metropolitan Area Planning Council, *Open Space and Recreation Study for the Boston Region* (1965).

[20] The Metropolitan Council of the Minneapolis–St. Paul area has twice blocked construction of a new major jetport 20 miles north of the city on the grounds of potential ecological damage. While the Twin Cities metropolitan agency is endowed with substantially more power than most such agencies elsewhere in the nation, the airport case is not definitive proof of environmental leadership. In the past decade airport construction has been blocked as a result of protests by affected local communities. The Twin Cities council does have extraordinary power, however, since the state legislature has given it the right to levy taxes, sell bonds, and suspend projects of single-purpose, functional agencies it finds inconsistent with its regional plan.

MAPC also attempted to provide a general guide for regional land development through a density umbrella concept. This seems to have had little regional impact on the communities or state agencies.

(4) *Change the system?* Perhaps the most sweeping reform legislation introduced in 1970—not by MAPC—was a bill filed calling for a constitutional amendment to create regional government in Massachusetts. The proposed legislation would not have abolished cities and towns, nor, in theory, would it have transferred powers now held by the state to a regional entity. In practice, however, a new governmental entity established at the regional level and endowed with genuine power would necessarily represent a transfer of power now held by other government units. In early 1970 the MAPC established a Technical Committee on Regional Organization and appeared to be moving in the direction of a full-fledged council of governments, with the establishment of a regional government far off on the horizon.

Subsequently, under the so-called Phase II of state agency reorganization, the state was subdivided into subregions of various sizes in the expectation that the legislature would approve plans for a substantial amount of regionalization of agency responsibilities. The legislature having proved unreceptive in its 1972 session to this proposal, the Greater Boston Chamber of Commerce advanced legislation creating a "middle-tier" metropolitan governmental body. Limited initially to certain functional areas, the new organization will, if appropriate legislation is enacted in the 1974 session, focus on a favorite area topic, solid waste disposal among its early priorities. Interestingly, the legislation provides for the absorption of the Metropolitan Area Planning Council as a research and advisory component of the new metropolitan agency.[21]

In effect, the Chamber proposal embodies some of the ideas advanced in the late 1960s by Mayor Kevin White of Boston, concepts which find support from some state agencies and state representatives. Based on the opinion of local observers, however, the prognosis for the "middle-tier" proposal is not at all good. A combination of a weak MAPC and administrative decentralization of state agency operations seemed much more probable toward the end of 1973.

Air Pollution

Air pollution occurs when the capacity of the air to dilute the pollutants is overburdened. Major pollutants are grouped into five main classes: carbon monoxide, hydrocarbons, particulates, sulfur oxides, and nitrogen oxides.

Not only does air pollution significantly affect human health—in particular, the incidence of such chronic respiratory ailments as emphysema and

---

[21] See Greater Boston Chamber of Commerce, *"TASK FORCE,"* pp. 27–31.

bronchitis—but it also causes costly damage to plant life, buildings, and materials, reduces visibility, and alters climatic conditions. Although it is difficult to assess the full costs of air pollution, the total probably amounts to billions of dollars annually.

> The cost of using the public skies and streams as dumping grounds for private sewage is appalling. Air pollution alone costs Americans about $18 billion a year in medical expenses, cleaning bills, and building maintenance. In poisoned New York City, the cost per family may reach as high as $620 per year per family, according to a study released about a year ago by the state air resources division.[22]

The first government legislation concerned exclusively with air pollution was passed in 1955, a federal bill authorizing $5 million for the Public Health Service in the Department of Health, Education, and Welfare for research, data collection, and technical assistance to state and local governments. Subsequent legislation passed in 1963 and 1965 provided federal enforcement authority to attack interstate air pollution problems and included national regulation of air pollution from motor vehicles. Legislation enacted in 1967 and 1970 also provided for a regional approach to establishing and enforcing federal-state air quality standards. With the exception of federal regulation of emissions from new motor vehicles, primary responsibility for the control of the sources of air pollution rested with state and local governments. State and local agencies, short on funds and qualified staff, have been slow to develop effective action programs, and their reluctance to take violators into court has led to a rather casual attitude toward enforcement.

To qualify for federal aid, states must set standards limiting the levels of pollutants and promulgate plans for implementing these standards, setting specific maximum emission levels by source and a timetable for achieving compliance. Air quality standards are based on federally determined minimum levels, which are subject to change. Legislation requires state and local governments to comply if they wish to continue to qualify for federal funds. Thus the federal government acts as the stimulus and guarantor of minimum national standards, leaving enforcement of these standards as a state responsibility.

Serious efforts aimed at enforcement and control of air pollution are underway throughout the nation. The primary impetus has been a combination of alarming publications and federal research, regulations, and funding, though some states confronted by especially serious problems—notably California—have taken vigorous action.

[22] Gurney Breckenfeld, "Environmental Protection Devices," *Planning 1970*, Selected Papers from the American Society of Planning Officials National Planning Conference (Chicago: American Society of Planning Officials, 1970), p. 218.

Massachusetts has not been a leader in this field. Despite the importance of the problem, air pollution, until the early 1970s, received only minimal governmental attention, most of it desultory enforcement by state or, more rarely, local officials of visually sighted chimney violations. In Massachusetts, air pollution has been defined as a public health problem, and responsibility for action has largely been assigned to the state Public Health Department. To be sure, the department's operating Environmental Health Division is assisted by an advisory council consisting of various state commissioners and eleven public members appointed by the governor, but the department runs the show, even to approving the formation and operation of area air pollution control districts[23] and to approving rules and regulations promulgated by local governments.

The state Department of Public Health has reacted to allegations of feeble and dilatory administration of air pollution legislation with the traditional response: legislative failure to provide adequate funding has left the Division of Environmental Health badly understaffed. Further, tougher regulation might increase unemployment either by forcing marginal companies out of business or by providing an incentive to relocate out of state. (Both arguments are also frequently used in explaining the need for gentle enforcement of water pollution and other environmental legislation.) At the municipal level, the City of Boston has a small staff responsible for smoke inspection and much the same failure at enforcement is in evidence.

The Registry of Motor Vehicles has played a lesser role in controlling air pollution. The registry, independently of the activities of the Department of Public Health, is involved in the problems of motor vehicle emissions. As new, stricter regulations regarding emission control devices and other efforts to lower auto emissions are developed, the registry will be concerned with their enforcement.

The existence of serious problems with major regional implications would appear to offer a great opportunity for action at the metropolitan level. At the very least, conferences, seminars, and publications could be mounted as part of a broad citizen-education campaign. In practice, however, air pollution has remained on the back burner. MAPC, to be specific, has chosen other areas on which to concentrate its limited funds and staff resources. Unless there is a fundamental change as the result of state reorganization, the federal government will be the key to progress in this area; the new automobile exhaust legislation may have far more impact than the corps of chimney-smoke inspectors employed by the state.

The pattern of research and lethargic action in the air pollution field was decisively broken in mid-1973 when the federal Environmental Protection

---

[23] *Compendium of Environmental Legislation*, p. 40.

Agency, through its regional offices, dropped a bombshell on many of the nation's metropolitan areas, including Boston. The background for the furor is sketched in the Federal Register, which summarized the agency's rulings for transportation control. Briefly, the proposed rule would require states to plan for sharp cutbacks in auto emissions in highly polluted areas, particularly central business districts, but also including other affected areas.[24] The difficulty is that a radical restructuring of urban transportation patterns and substantial investments would be required to achieve a major reduction in such emissions, as for example, banning automobiles from sizable portions of central cities. More stringent enforcement of existing control mechanisms would be helpful but insufficient to meet the EPA's 1975 standards.[25] The special air pollution problems affecting the Boston area were discussed in another EPA-funded study published in mid-1973, which called for a broad action program, including restrictions on parking and movement of "nonessential vehicles" in and around the City of Boston to attain the EPA standards for hydrocarbon and carbon monoxide emissions.[26]

The reactions have been predictable; in Boston as elsewhere businessmen and politicians have reacted violently in the negative, and environmental agencies have attacked the proposals as premature, overly restrictive, and unrealistic. The upshot has been an EPA retreat, with state and local agencies in some disarray, torn among medical evidence, their interpretation of political realities, and the prospect of recurring federal pressure. Many officials seem to hope that the problem will somehow disappear as transit lines are constructed and a rotary engine or some other technological advance comes to the rescue.

Noise Pollution

It is generally agreed that prolonged exposure to excessive noise levels (over 85 decibels) can cause serious physical and mental problems. Industrial safety regulations call for suppression measures and, where these are not feasible, require earmuffs to be worn by workers in extreme situations, such as ground crews in proximity to jet planes during landings and take-offs. The

[24] Environmental Protection Agency, "Requirements for Preparation, Adoption, and Submittal of Implementation Plans, Proposed Transportation Control Measures," *Federal Register*, vol. 38, no. 3, January 12, 1973.

[25] *Evaluating Transportation Controls to Reduce Motor Vehicle Emissions in Major Metropolitan Areas*, prepared by Institute of Public Administration and Teknekron, Inc. in cooperation with TRW, Inc., for the Environmental Protection Agency, Office of Air Quality Planning and Standards, Research Triangle Park, North Carolina, November 1972. See pp. 8–12, 17–19.

[26] Wallace Woo, Salvatore Bellomo, and John Calcagni, "Strategies to Reduce Air Pollution: The Boston Experience," paper presented to the 66th annual meeting of the Air Pollution Control Association, Chicago, Illinois, June 24–28, 1973.

far more prosaic environments created by high-volume rock bands, construction, heavy traffic, and even household appliances can also cause loss of hearing.

The adverse effects of noise are not yet well understood. Those generally recognized include hearing loss, disruption of normal activity (classes in school buildings located near airports are often interrupted while a plane taking off or landing passes close overhead), and general annoyance. Loud noises, such as the sonic boom, can even cause physical damage to structures. The most common and best understood physiological effect of noise is permanent hearing impairment. This occurs, as a rule, only under industrial conditions, although there is evidence that prolonged exposure to rock bands, outboard engines, or snowmobiles can damage hearing. Temporary physiological changes can be produced by noise—for example, constriction of the smaller arteries, which results in increased pulse and respiration rates. Nevertheless, much more research is needed to understand this phenomenon fully. Determining psychological effects of noise is more difficult, because individuals vary widely in psychological sensitivity to noise; what is annoying to one person may not be so to another. As a quick guide to determining the psychological safety of noise, Aram Glorig, a noted researcher in this field, suggests that when a person is in the presence of noise loud enough to require that person to raise his voice in order to be heard, the noise is at a potentially harmful level.

The greatest contributor to high noise levels in the urban environment is transportation: trucks, buses, motorcycles, airplanes, and rail systems. Sustained traffic noises in today's cities on the surface or in subways can reach 90 decibels, and it is generally agreed that prolonged exposure to levels of this sort can cause permanent hearing loss. Airports are another major source of noise pollution: a four-engine jet generates 115 to 120 decibels at take-off. Heavy doses of these noise levels from airports built and quickly expanded close to major urban areas have produced many citizen complaints and even lawsuits, and will produce more. The American Operations Council International estimates that, by 1975, "15 million people will be living near enough to airports to be subjected to intense aircraft noise."[27]

Efforts to enforce noise pollution regulations must be concentrated primarily on the local level. The federal government, however, has already taken steps to control certain noise conditions: in 1969 occupational exposure and noise standards relating to aircraft noise were established. The federal industrial standard calls for a maximum exposure of 90 decibels for an eight-hour period.

[27]*Environmental Quality*, pp. 124–126.

State and local governments have yet to take a significant role in this area. Many states appear to be waiting for the federal government to lead the way and propose guidelines and standards that the states could follow. However, both Colorado and California have enacted noise control legislation much more stringent than federal legislation.

In Massachusetts, legal responsibility for noise control rests with the Department of Public Health's Division of Environmental Health. The department's concern here, as in the cases of air and water pollution, is primarily with the health aspects of the problem; namely, setting acceptable noise standards and regulations for maintaining and enforcing those standards. Both tasks are still in the embryonic stage. Regulations for the control of air pollution established by DPH, effective July 1, 1970, also declare under Regulation 10 that

> no person owning, leasing, or controlling a source of noise shall willfully, negligently, or through failure to provide necessary equipment, service, or maintenance or to take necessary precautions cause, suffer, allow, or permit unnecessary emissions from said source of noise.[28]

Such regulations are almost meaningless, and difficult or almost impossible to enforce. (A largely overlooked repository of power to deal with excessive noise is local nuisance law, which generally prohibits individuals and corporations from imposing hardships on their neighbors. To date, local communities have not been willing to enforce these regulations.)

Noise abatement is also the concern of the transportation agencies—the Registry of Motor Vehicles and the Massachusetts Port Authority. The registry receives complaints about motor vehicle noise; the Port Authority cooperates with the FAA in ensuring that aircraft comply with FAA regulations on take-off and landing noise.

However, at least two branches of state government (along with the City of Boston), the Port Authority and the University of Massachusetts, have been involved in one part of the noise problem—flights over Logan Airport. One flight path crosses over East Boston, and the airport has been a major source of complaints in this section of the city. A consultant study provided guidance for planning and redevelopment activities in relation to airport noise patterns.[29] The University of Massachusetts, whose new urban campus, under construction at Columbia Point, is directly under a major flight path, has given considerable attention in its building program to the problem of recon-

---

[28] Massachusetts, Department of Public Health, *Regulations for the Control of Air Pollution* (1970), Reg. 10.1.

[29] Bolt, Beranek and Newman, *Aircraft Criteria for University of Massachusetts, Boston Campus at Columbia Point*, prepared for Pietro Belluschi and Sasaki, Dawson, and DeMay, Watertown, Mass. *Technical Report 1919*, February 1970.

ciling teaching, study, and other university activities with frequent and loud bursts of noise.

From the standpoint of a metropolitan agency casting about for neglected areas in which there is insufficient research and planning, noise pollution would seem to offer exciting possibilities, even though most of the action on this front is taking place in federal agencies. Far too little is known concerning the problem and its solution, but the Boston area does contain a number of eminent authorities in the acoustical field. In theory, noise pollution would appear a promising field for research, convening conferences, etc. In practice, however, the MAPC publication *Guides for Progress* makes no specific mention of the noise problem at all in its discussion, "The Quality of the Physical Environment," presented in the agency's 1968 guide plan.

Solid Wastes

Rubbish disposal has been a municipal responsibility almost from the dawn of recorded history. Although collection and disposal are now regarded as problems from a cost-benefit standpoint, in earlier times the value of garbage, compost, and fill more than outweighed the costs of assembling and eliminating solid waste products. Even in modern times, developing nations with a good deal of tolerance for odors and disorder, much manpower, and few viable alternatives can make effective use of waste products. It is primarily in highly developed nations that the removal of solid waste is approaching crisis proportions. Interestingly, despite long, detailed municipal experience, much remains unknown regarding the potential technology and potential benefits accruing from reuse of solid waste products.

There are some who fear that America will be suffocated in a rising ocean of "solid wastes," a euphemism that covers a vast range of throwaway materials of which paper, metals, rubber, and glass are the most prominent. At an estimated 1 ton annually per capita, the nation's urban areas now generate about 250 million tons of solid wastes a year, which are disposed of at a cost of over $4 billion a year. This works out to $21.50 per capita per year for disposing of almost a ton of publicly collected solid waste per capita per year.[30] The various methods of disposal cost somewhat differently:

> Landfill costs amount to only one to three dollars per ton, compared for example to incineration, where costs run from about three to ten dollars per ton, depending on the size of the installation and whether some use is made of the heat generated. Increasingly sophisticated stack gas cleaning equipment, dictated by stricter air pollution standards, can be expected to

---

[30] American Public Works Association, *Municipal Refuse Disposal* (Danville, Ill.: Interstate Publishers, 1970), p. 10. The APWA estimates that the actual per capita figure is 10 pounds a day but only about half is collected by public agencies (ibid., p. 25).

raise these costs still further. Large-scale composting, in the few places where it is done in the U.S., has a comparable price tag, about five to ten dollars per ton, minus whatever credit can be obtained from the sale of compost and other reclaimed materials.[31]

In Massachusetts, solid waste disposal is primarily a municipal responsibility, subject to approval by the state Department of Public Health. The state agency is also required to approve local plans for the formation of regional disposal districts.[32]

A late claimant in the solid waste field is the state Department of Public Works. In 1969, the road-building agency was authorized to establish a Bureau of Solid Waste Disposal. In cooperation with the state Department of Public Health, the DPW was authorized in 1969 to survey needs, hold hearings, and devise proposed program and implementation plans for solid waste disposal.[33] The DPW can also acquire land and structures but must secure approval from the state Departments of Public Health and Natural Resources for such operations and must submit its projected budget for approval to the included cities and towns. Additional legislation enlarged the powers of the Department of Public Health, in requiring its approval of sites, proposed use, and plans and design of facilities operated by other agencies of the commonwealth.

This legislation involving the allocation of responsibilities between the DPW and state Department of Public Health relates to an internecine power struggle involving MAPC and MDC on one side and the DPW—the victor—on the other. Early in its planning, MAPC selected solid waste as a handy target of opportunity for its research and development activities. In a three-volume report issued in 1968, MAPC asserted that by 1970 most of the metropolitan area's open dumps and landfill facilities would be exhausted and proposed a comprehensive regional solid waste disposal program assigning implementation responsibility to the MDC. Highlights of the proposal included:

(1) Construction of nine regional incinerators and eight sanitary landfill operations in order to dispose of all domestic, commercial, and industrial solid waste generated in the metropolitan area through 1990.

(2) Total capital improvements of an expenditure of $85,000,000 to be raised by general obligation bonds. A $59,000,000 bond issue should be authorized in 1967; an additional $15,000,000 bond issue will be needed by 1973; a third $10,000,000 bond issue will be needed by 1984.

[31] Robert R. Grinstead, "The New Resource," *Environment* 12, no. 10 (December 1970): 4.

[32] *Compendium of Environmental Legislation*, pp. 36–37.

[33] Massachusetts General Laws, 22: 18 (1969).

(3) Total expenditure for incinerator construction through 1990 of $47,600,000 at current prices. This estimate assumes a high level of design utilizing modern technological devices to meet the most stringent air pollution standards.

(4) A statewide assistance program to stimulate high quality service throughout the Commonwealth. This program should provide grants to regional disposal agencies for a portion of the capital costs of solid waste facilities, including incinerators, sanitary landfills, and other approved disposal methods.[34]

The MAPC proposal elicited a swift reaction. Although the highway agency had only tangential connection with solid waste disposal programs, the DPW quickly formulated an alternative course of action, in which it rather than the MDC would be the dominant party. As a summary of the legislation suggests, the DPW rather than the MDC-MAPC combination garnered the legislative votes. Subsequent legislative reservations concerning the DPW resulted in further legislation giving the state Department of Public Health veto power over the DPW's principal activities in the solid waste field. However, the MAPC initiative had yielded no results for itself or the MDC, the other partner in the 1968 solid waste study.

It may be noted that a 1972 review of metropolitan solid waste problems conducted by the DPW concluded that

. . . inter-community cooperation in meeting solid waste disposal needs had been non-existent since the (1968) MAPC report was issued and that the crisis was worsened by the continued obsolescence of present incinerators and by more stringent particulate emission levels issued by the state Department of Public Health. . . .[35]

Solid waste lends itself more readily than most environmental concerns to metropolitan planning and program implementation. The technical requirements of collection and transportation dictate metropolitan or regional solid waste disposal solutions rather than the more limited, overlapping, and less efficient muncipal programs. Large areas—with 500,000 population as a possible optimum size—are suggested as the basis for planning disposal systems. A regional role still appears natural in this area. Certainly, there is a useful precedent for metropolitan action on the solid waste problem. As in other

---

[34] Metropolitan Area Planning Council, "Summary," *Solid Waste Disposal for Metropolitan Boston* (review draft), vol. 1, p. 12. The study was prepared in cooperation with the Metropolitan District Commission and the Department of Public Health, and was financially aided by HUD Section 701 funds.

[35] Greater Boston Chamber of Commerce, *"TASK FORCE,"* p. 6, quoting *Raytheon Service Company Report*, prepared for the Bureau of Solid Waste Disposal, Massachusetts Department of Public Works, May 1972.

aspects of metropolitanism, the Twin Cities have led the way, in Norman Beckman's words, by

> building on the already unique quasi-governmental Twin Cities Council by adding operational responsibilities for parks and solid waste disposal. This action by the 1968 Minnesota legislature points a way to an evolutionary expansion of metropolitan councils of government into true governmental entities.[36]

Esthetics

No more than a generation ago there was widespread indifference to environmental esthetics, an attitude inherited from both the frontier and the early industrial revolution—the linkage between "muck" and money characteristic of the gloomy English midland cities. Ugliness was not permitted to diminish civic pride. In the author's personal experience in the mid-1950s, an outraged foreigner who described a decaying New England textile city as a "rubbish heap" was soon thereafter invited to quit the city's employ. Some cities and neighborhoods apparently engendered a reverse sentimentality that gloried in the tough, character-building aspects of the stench and filth. Some rhapsodized about New York's East Side, others praised the West End or South Boston, Philadelphia's Kensington, or Chicago's North Side as places which may have appeared brutally unattractive to outsiders but which throbbed with neighborliness and charm for the insider.[37]

Furthermore, smokestacks belching great clouds of pollutants into the air were hailed as a visible sign of progress and payrolls. People were happy to have the paychecks that resulted from this prosperity. Only much later, when the standard of living had risen beyond the subsistence level, did people begin to concern themselves with the esthetics of their communities as well as the economic condition.

Of late there has been a growing interest in environmental esthetics. The quality of some institutional architecture seems to be improving: office buildings, churches, city halls, airport terminals, and even some residential structures have undoubtedly been upgraded in the past two decades. The least progress seems to be evident in low- and middle-income housing. All too often housing seems to be constructed without an architect or, at best, with a low-priced hack who enlarges his net profit by using shopworn drawings from the files. Although good architecture need not add to costs, developers tend to regard it as a frill and to view architects as impractical prima donnas more

---

[36] Norman Beckman, "Legislative Review—1968-1969, Planning and Urban Development," *Journal of the American Institute of Planners* 36, no. 5 (September 1970): 346.

[37] For an eloquent exposition of this view, see Jane Jacobs, *The Death and Life of Great American Cities* (New York: Random House, 1961).

interested in erecting monuments to their own egos than in building usable buildings in the real world. Too often there is more than a grain of truth in the suspicion that what many architects really desire is a patron like the Ford Foundation, Seagrams, or Lorenzo the Magnificent, not a struggling developer hewing close to FHA specifications. First-rate architects, on the other hand, are fully aware that every substantial building is a public monument good for a century or more, in which it either blights or enhances its environment. Amortized over 50 to 100 years, even a sizable architect's fee and an additional 10 percent added to construction costs seem rather modest.

One of the chief obstacles to stronger public commitment to urban esthetics is the difficulty of representing environmental beauty on the balance sheet. It is often less expensive to avert one's eyes and reserve environmental appreciation for trips to the country or European junkets than to spend the money to ensure a good urban environment. Political decision makers and businessmen have been slow to see that a high-quality environment offers a priceless competitive advantage. In an earlier era this was perhaps less important. The jobs were in Scranton, Duluth, Pittsburgh, Gary, Cleveland, and Buffalo because that was where resources were processed. In an age of footloose industry and mobile executives, however, the urban area that offers attractive living conditions for the sophisticated has a distinct advantage. Advertising in business magazines places heavy stress on environmental amenities. Firms located in San Francisco, Boston, Atlanta, Denver, Madison, or Ann Arbor find it easier to attract first-rate professional, scientific, and academic talent and thus these areas attract technologically advanced industries.

Among architects and city planners, the Boston area is considered one of the most attractive parts of the nation. It has a combination of what many areas lack, in whole or in part: a strong central focus, a rich inheritance from the past, beautiful natural settings, a sense of human scale as compared to an enormous aggregation like New York, and proximity to a variety of high-quality recreation areas. Some of the area's inherited esthetic values are attributable to government action—some municipal, like the Boston Common and Public Garden, and others regional, like the MDC park and parkway system. The preservation of so large a proportion of nineteenth-century structures is a matter of good fortune. The area's relatively slow economic and population growth since the early 1900s, when an enormous amount of destruction of attractive older buildings occurred in fast-growing areas, resulted in a retention of both charming external facades and neighborhood cohesiveness to an extent not found in many other parts of the nation. Nevertheless, the area suffered brutal blows, particularly in the 1920s and beyond, from callous subdividers, outdoor advertisers, and perpetrators of much insensitive architecture. Not until the late 1950s was there much evidence of governmental action to preserve and improve the esthetic environment. Legislation enacted during the 1950s included historic preservation acts, the creation of a powerful, design-conscious redevelopment agency in the City of Boston, and wetland and open space legislation, which helped to preserve and extend some of the area's greenery. Nevertheless, progress has been spotty, more a matter of erecting individual structures like office buildings, residences, churches, and schools than of upgrading the esthetic qualities of entire neighborhoods and communities. The conscientious step, stoop, and sidewalk scrubbing in working-class Baltimore or Philadelphia neighborhoods (and in such nations as Holland) has never been part of the life style in Boston or, indeed, in most other cities.

At the state level action on the esthetic front has been confined to hiring good architects to erect some state office and education buildings. Control of outdoor advertising has been feeble, to say the least, but individual agencies have done some remarkable things: the newly remodeled MBTA stations have deservedly won prizes for good design and some of the Port Authority's airport buildings are outstanding examples of good architecture.

Despite Boston's efforts with its outstanding City Hall and the design review procedures of its redevelopment authority, local government activities in this field have been limited. A number of communities have acted to control outdoor advertising, to preserve attractive open areas, and to ensure that new subdivisions will have adequate parks and trees, and some have established historic districts to preserve attractive older buildings. But there have been no community-wide programs in urbanized communities disfigured in the past decades to upgrade esthetic quality, to deal with the backlog of inherited problems that turn whole streets and neighborhoods into collections of eyesores.

At the metropolitan level less has actually occurred, although much has been proposed. The MAPC's chief thrust in this field has been through its open space proposals. Prepared by architectural landscape consultants, the proposals seek to enhance the region's natural assets while meeting its needs for recreation and public services. The final proposal was a composite plan combining the 200,000 acres of existing wetlands, water, and other open space in the region[38] with a secondary system of wedges and greenbelts. One objective was to give form and definition to the region by channeling future urban development into seven corridors radiating from the core area and separated by six wedges of relatively undeveloped land.

The open space study, prepared in cooperation with the MDC and the DNR, incorporates the plans and aspirations of these two agencies in its proposal and accepts basic plans of the transportation agencies as givens. How-

---

[38] Metropolitan Area Planning Council, *Open Space and Recreation Plan and Program for Metropolitan Boston*, vol. 1, pp. 12–17.

ever, implementing the plan involves the expenditure of over $200 million at 1967-1968 price levels. In an era of rapidly rising prices and constricted state budgets, progress in acquiring open space has been painfully slow. Furthermore, the plan involves acquiescence, if not overt approval, by a number of other actors, including state agencies, municipalities, and private developers. Aside from the nagging shortage of ready cash, in the early 1970s there was no sign that the plan had achieved the kind of consensual, sacrosanct status that would lead to basic alterations in other plans and programs needed for effectuation. The plan had, in fact, gone out of print (and perhaps out of mind) while the MAPC devoted its energies to other topics.

Conclusions

This brief examination of environmental problems and approaches in metropolitan Boston leads to several preliminary conclusions:

(1) In small compact states in which the largest city is also the state capital, the distinction between state and metropolitan approaches may become blurred. State action via line agencies or quasi-regional agencies may be the most effective and desirable approach to many urban problems, including those relating to the environment.

(2) A regionally based agency risks failure when it attempts to organize and implement programs involving the awarding of large contracts. In running down the list of potential areas for attention, the MAPC naturally selected research in population, land use, and economic base, the planning profession's trinity. Going beyond this basic research, the metropolitan agency focused on several major areas in which substantial research was obviously needed and in which technical studies appeared appropriate to provide expert answers to technical problems. MAPC concentrated much of its attention on technical areas that seemed to offer the advantages of an open turf unoccupied by unfriendly agencies and that promised broad public acceptance. In none of these cases, the open space study, the airport study, or the solid waste study, did there appear to be powerful built-in resistance to MAPC proposals from established line agencies.[39] In each case MAPC's initial calculations proved totally erroneous: rival occupants of the turf on which MAPC had impinged

---

[39] An exception to MAPC policy of avoiding open confrontation occurred in early 1970, when the governor was reported to have blocked publication of a MAPC staff report that attacked his proposed $3.5 million restudy of Boston area highways on the ground that further study would cause unconscionable delays in construction of vitally needed roads. Gubernatorial staff personally phoned members of the MAPC transportation committee to head off their endorsement (*The Boston Globe*, February 5, 1971, p. 6).

leaped into action and in the end all of the proposals came to nothing.[40] What appeared to be barren tundra sprang into luxuriant growth after only a light sprinkling from MAPC's watering can.

(3) An independent regional agency can (but in the case of MAPC did not) play an effective role in educating and guiding public opinion and decision-maker attitudes. The private query of a prominent academic-political figure as to whether MAPC was "dead or just moribund" is revealing. MAPC seems largely to have faded from public view in recent years despite the opportunity for breaking new ground by focusing on attention-getting issues like the environment. (It may be noted that this need not be costly. Seminars, publications, and conferences can be arranged at little expense.)

(4) The importance of personalities should not be underestimated. The metropolitan agency enters an already crowded arena, and necessarily encounters difficulties in second-guessing or otherwise interacting with suspicious line agencies. Breaking into a closed circle calls for luck, tact, or ruthlessness—perhaps all three.

Robert C. Wood describes the metropolitan area as consisting of "an alienated and indifferent public . . . highly diverse and unstructural set of elites . . . highly decentralized and highly volatile patterns of influence."[41] Over a decade ago, Wood suggested that the planner play a politically activist role, fashioning a strategy to attract the support of diverse interest groups, and building viable coalitions to implement plans and programs. But the metropolitan planner is in an extremely weak position. He has no serious political leverage:

> without the kind of power base that could be provided by metropolitan government, or perhaps by an aggressive governor, he has two alternatives: long-term public and agency education for planning, which may pay off in five, ten, or twenty years, and/or the kind of ritualistic charades which Charles Adrian maintains is all that the metropolitan areas really want. Coalition-building in the metropolitan jungle is clearly no pathway to short-term or even longer-term injection of comprehensive planning in metropolitan areas.
>
> Implementing metropolitan planning on an advisory basis calls for the qualities of a near-genius who is blessed with extraordinary luck. This combination seems to be in extremely short supply.[42]

[40] In this respect an observation of Edward Banfield's may be helpful: metropolitan planning can lead to nothing because there is no possibility of agreement on the general interest (Edward C. Banfield, "The Political Implications of Metropolitan Growth," *Daedalus* 1 [Winter 1961]: 72). It is paradoxical that Banfield's comment was first delivered at a seminar on metropolitan planning at the Joint Center for Urban Studies sponsored by MAPC.

[41] Wood, *1400 Governments*, p. 114.

[42] Melvin R. Levin, *Community and Regional Planning: Issues in Public Policy* (New York: Praeger, 1971), pp. 142–143.

Directions for the Future

Organizational Structure

The analysis of environmental problems and governmental response in metropolitan Boston suggests that the area is typical of most SMSAs in four respects: (1) the problems have outrun the existing agency structure; (2) relatively little is being done by locally controlled agencies (i.e., the municipalities and MAPC); (3) there is a substantial impact on the metropolitan environment through the activities of state agencies, especially those that are metropolitanized—the MDC, MBTA, the Port Authority, and the Turnpike Authority; and (4) key state agencies, particularly the DPW and the Departments of Natural Resources and Public Health, are playing significant roles in shaping the environment and attempting to regulate some of its abuses. But Boston's central city constitutes an extraordinarily small proportion—less than 25 percent—of SMSA population, compared to 40 percent or higher in other U.S. metropolitan areas. In a direct sense, metropolitanism is a compensatory mechanism for the premature ossification of central-city boundaries which has occurred in virtually all metropolitan areas. Since, with few exceptions, the central city could not continue the organic process of growth by accretion into the present century, metropolitanism offers a way of reconciling geography and problems. The question is how best to go about it.

State government. Since the powers to regulate and fund environmental programs are focused in state government, it may be more realistic to modify the present arrangement than to attempt a radical restructuring that involves the very difficult problem of persuading the state to cede some of its prerogatives to another level of government. This does not rule out a significant restructuring of state services on a metropolitan basis.

One alternative approach to the problem of metropolitan governance involves state action to merge the metropolitan giants—the MAPC, the MDC, MBTA, and the Turnpike and Port Authorities—and to add to the new multipurpose agency the functions of the DNR in open space, of the Department of Public Health in air and water pollution, and of the DPW in solid waste disposal. The new agency would therefore be responsible for most of the transportation functions in the area, most sewer and water systems, parkways, and regulation of environmental problems. This would not diminish state power, but would regroup key executive agencies to provide delivery of services and regulatory operations on a metropolitan basis. Although citizen advisory boards and/or advisory and review boards drawn from local governmental officials might be desirable, the multipurpose agency would not have a governing board appointed by the governor on the Twin Cities model or an elected board on the Nashville model. It would represent a new approach to state government rather than a metropolitanization of power. The combination of Phase I and the proposed Phase II of state government reorganization goes a considerable distance in this direction.

There are two reasons for adopting this approach:

(1) Difficult as it may be to effect a transfer of power within the state executive agency structure, it is probably much less difficult than persuading state government to divest itself of many of its powers.

(2) Citizen representation (particularly in a small, compact state where the state capital is also the chief regional city) can be adequately assured through the regular process of state government. Although a new, autonomous level of government may be needed in large sprawling states with a sizable urban-rural cleavage, the case is much weaker for states like Massachusetts, Rhode Island, Connecticut, Delaware, Maryland, and perhaps New Jersey, where state government already approximates genuine urban government.

So far as environmental problems are concerned, some are clearly metropolitan in nature or could well be tackled on a metropolitan basis by state government. All this, of course, does not rule out a substantial role for ombudsmen and public interest law firms.

Metropolitan government. In late November 1969, Mayor Kevin White of Boston proposed creation of a strong council of governments (COG) consisting of 100 municipalities in Eastern Massachusetts. Unlike a typical COG (a voluntary association of elected officials but a form of metropolitan government), the new metropolitan agency would include the MAPC, MBTA, Port Authority, MDC, and Air Pollution Control District, an assemblage of agencies closely resembling the list recommended above in the proposed state agency metropolitanization. The difference lies in the transfer of power from state government to elected officials of the cities and towns. That this approach is politically infeasible (as well as subject to attack on other grounds) is demonstrated by the cold reception accorded to Mayor White's proposal: the bill introduced in the 1970 legislative session went nowhere. As noted earlier, the proposal was due for resubmission in the 1974 session, but its prospects are not much brighter than in 1969.

Proposals to reorganize government to deal with metropolitan problems must take cognizance of Robert Wood's observations about the New York region. In the late 1950s, he saw the mood of the citizenry, elected officials, and agency heads as a blend of some discontent, considerable satisfaction, and a good deal of apathy. Certainly Wood has been proved correct in his prediction that no drastic reorganization of government was on the near horizon.[43] Perhaps apathy and contentment have diminished and dissatisfaction has grown in the 1970s. The climate for agency reorganization, at least, has improved, if only because of rising standards of education, incomes, and

---

[43] See Wood, *1400 Governments*, chap. 5.

expectations. Existing and potential constituencies for an approach to metropolitanization are slightly stronger and more important and the opposition weaker or at least more apathetic than in the fifties and early sixties. But while the time may be ripe for a consolidation and metropolitanization of state agency functions, I doubt that state government has decayed to the point of committing self-castration through ceding power over its largest and richest region.

Some Topics for Research

The preceding analysis points to a number of research priorities, some directly related to metropolitan impacts and programs linked to environmental problems, and others addressing unanswered technical questions. Five areas have been tentatively identified as warranting further study.

(1) *Field-tested models.* Areas that have developed successful administrative and technical responses to specific environmental problems should be carefully examined. These include relevant, successful experience at the federal, state, and municipal, as well as the metropolitan level, and foreign and nongovernmental programs. These models should be studied for wider applicability through metropolitan or other appropriate vehicles. Where indicated, useful domestic or foreign experience should be translated and converted to permit broader adoption and application. The Twin Cities, Toronto, Dade County, Nashville, and other domestic and foreign examples offer valid subjects for study and possible emulation.

(2) *Interfaces.* The border areas where governmental responsibilities mesh (and/or overlap), where environmental problems merge into wider considerations of urban and national policy and goals, and where economics, changing life styles, the quality of life, and political realities converge all require investigation. For example, until convincing answers are forthcoming on the relationship between economic growth and environmental protection (i.e., how a reconciliation or balance can be achieved), there is unlikely to be a major shift in government policy. As is the case with any important issue, consideration of interrelationships inevitably leads to an attempt to place the problem in its broader perspective. Consideration of this perspective represents an urgent, massive, and high-priority research item.

(3) *Costs and benefits.* Far too little hard data is available on costs and benefits involved in (a) continuing along present lines or (b) modifying current trends. Data on air and water pollution, noise pollution, and land use remain too fragmentary and inadequate to provide either policy guidelines or ammunition for public education and persuasion of policy makers.

(4) *Technological considerations.* Far too little is known concerning improved techniques of air and water pollution control and noise control, among other matters. Control of emission sources, screening out unfavorable environmental influences in dwellings and other structures, inexpensive, rapid testing techniques for new materials, and reuse of waste materials all deserve urgent study. Another area for research is the extent of unintended consequences; solutions for one problem may create other problems. An example is the automobile emission control device, which reportedly aggravates smog problems.

(5) *Staffing needs.* One of the critical problems confronting every level of government, and particularly acute in newer programs, like environmental efforts, is ensuring an adequate and effective staff. Enlarged environmental efforts will require substantial numbers of trained professionals. High priority should be given to determining the number, characteristics, and training required of the administrative and technical staff needed to design and implement environment programs. Such inquiry might well be part of a larger study into governmental manpower needs, particularly in state and local governments.

4 A British Approach to the Reform of Metropolitan Governance: The Redcliffe - Maud Report 1966 - 1969

WILLIAM LETWIN*

Principles in Conflict: Efficiency v. Self-Determination

On Comparing Political Systems

English local government seems, on the surface, very different from American local government. Lacking states, England has a unitary rather than a federal system of government. Lacking a written constitution, it has a central government formally unrestricted in scope. Parliament is not barred, according to the prevailing theory of the English constitution, from doing anything it decides; for example, it can make whatever changes it may see fit in the structure of local government. Aside from the question of political feasibility, no explicit constitutional obstacle prevents Parliament from redistricting local government, reducing or enlarging its powers, or altering its organization in any other way. Neither does tradition erect any overwhelming obstacle to such far-reaching changes, since national rather than local government was the original source of sovereign power: for nearly a thousand years local governments in England have derived their powers from the exclusive sovereign at the center—in the early days from royal charter, later from parliamentary

*Reader in Political Science, London School of Economics, and Visiting Professor, The Sloan School of Management, Massachusetts Institute of Technology.

enactment. In America, sovereign states preceded the national government and surrendered to it only part of their sovereign powers. Relations between central and local government seem so different from those in England that one well might wonder whether either country could learn much from the other's experience.

Nevertheless, it would be misleading to exaggerate the undeniable differences between the two frameworks of government. Different as they are, not only in detail but also in deeper constitutional character, they are fundamentally similar inasmuch as they rely on the same underlying principles. Both are democratic, and in the same spirit, and because of that both share the same basic attitudes toward local government, even though their institutions are so unlike.

In any democratic and individualistic society, where government proceeds through representation and a multiplicity of parties, the forms of local government must recognize the interplay of two conflicting principles—self-determination and efficiency. According to the principle of self-determination, the best government is that which comes closest to making the same choices as each individual would have made had he been deciding by and for himself. According to the principle of efficiency, the best government is that which does its work—whether enforcing rules or providing services—at the lowest possible cost per unit of output. The difficulty that faces any system of local government within a liberal democratic state is that the two principles typically point in opposite directions. Self-determination tends, all else being equal, to reach a maximum in very small communities, whereas efficiency tends, all else being equal, to reach its peak in rather large communities. The principle of self-determination therefore presses for much local government, and that very localized, whereas the principle of efficiency presses for extended units of government, metropolitan rather than municipal, possibly regional rather than metropolitan, perhaps even exclusively national. Yet this tension must be resolved somehow when men set out to reform their institutions of local government and to consider how they might improve the governance of existing institutions. Although it would not be very useful for Americans to consider adapting English institutions to serve within a sharply different political civilization, it can be immediately instructive to see how Englishmen have recently tried to resolve this conflict between the urge to keep units of local government small for the sake of self-determination and the urge to make them larger for the sake of efficiency. Before we examine the English experience, however, it is worth exploring further the abstract relationship of the two principles.

On Self-Determination

The principle of local autonomy derives from the still deeper principle of personal freedom. All democratic communities endorse the rights of a man to be ascetic or spendthrift, to be withdrawn or sociable, to practice any religion or none, to follow whatever pattern of consumption appeals to his taste, and, more generally, the right to behave as he himself sees fit even while others differ from him or disagree with him. If each citizen enjoys these rights, it follows that a communal collection of such persons should enjoy them as a group. If *all* members of a community desire the bars to be open on Sunday or the schools to be closed on Columbus Day, their water to be fluorinated, chlorinated, and filtered, or their parks to be planted exclusively with heather–then, in a democratic society, there can be no question about their collective right to make those choices and to give them effect in the budgets and even the ordinances of that community. Self-determination by a local community thus derives directly from the principle of personal liberty, and it derives comfortably in the hypothetical instance of absolute unanimity.

When members of a community do not agree absolutely, politics and government must come into play. Such devices as representation and majority rule are adopted to reconcile, compromise, and suppress disagreement in order that a single, unified decision can be elicited from the welter of discordant views that may be held about any real problem in any real community. But despite this complication of procedure, the principle still stands that each community, each group of individuals, enjoys the right, as a community, to govern itself, a right derived from the right of each of its members to determine the conditions of his own life. And the goodness of the communal choice, in a liberal democratic community, must be measured by the criterion of how closely it conforms to the wishes of the individual members.

The likelihood that self-determination will succeed in satisfying the choices of *all* members of the community (and not merely of a bare majority) depends not only on the degree of consensus but also on the size of the community. People with similar views cluster together, partly by choice, partly because living together makes them feel and think alike. As a result, the smaller the area defined as an autonomous unit of government, the greater the likelihood that the community will tend toward substantial consensus about basic questions of public policy.

The point can be illustrated most readily, if not most pertinently, by considering the linguistic situation of India. Some hundreds of languages are spoken in India as a whole; most Indians speak only one of them; and those who speak any given language live next door to others who speak the same, for otherwise they could not speak at all. If the whole nation democratically chooses a single official language, then this must be either the plurality language, Hindi, which is the most common but is still the language of a small minority, or it must be a convenient international language, such as English. If, however, an official language is chosen separately by each state, then each will have a different language, in some cases even a language used by a *majority* of its inhabitants. And if the official language is determined separately by each city, town, and locality, then in practically every instance the local official language will be the language used *almost unanimously* by all inhabitants. So the approach to unanimity is highest in the smallest communities.

If anyone doubts that this illustration has any real relevance to American government, he need only consider the geographic agglomeration of religious sects, of people with similar economic interests, of citizens distinct from each other because of color, race, national origins, occupation, or style of life. On every one of these accounts, the possibility that a communal choice will satisfy the desires, dispositions, and tastes of *each* individual in the community increases as the community becomes smaller in its geographic extent and its population.

Ideally, from the standpoint of the principle of self-determination, units of government should be very small. The ultimate ideal would be realized only if each individual governed himself alone, because only then would the choices of government correspond perfectly to the choices of the governed. Needless to say, this ultimate ideal contradicts the essence of society. Men do live together with others, and they need or wish to live together despite the differences between them. To put it more exactly, the differences between them are precisely what would make society desirable and useful, even if it were not inevitable and natural. Government exists because men who disagree nonetheless live together and some of their disagreements must be overcome by imposing common rules. To postulate that each individual should, ideally, govern himself is therefore to postulate that society is undesirable and government, in the ordinary sense, unnecessary.

Nevertheless, though we reject the absurd extreme to which self-determination of local communities may be driven, this principle clearly is an inescapable corollary of freedom and democracy, and it points out the advantages to be had from allowing small communities to govern themselves.

On Efficiency

Self-determination is the power to choose what one wishes. Efficiency, on the other hand, is the power to get what one chooses, within the limitations imposed by the means available. Efficiency is regularly thought to be a function of size, and in some degree it is. If a village of a few dozen people were to install its own water-purification plant, the cost per gallon would almost certainly be higher than if the village were to buy purified water from a neighboring city whose plant was capable of treating millions of gallons rather than hundreds of gallons a day. For all the basic reasons that explain economies of scale in the production of any other good or service—specialization of labor and capital goods and the overcoming of lumpiness in the factors of production—public services rendered by the smallest units of government are almost certainly more costly than the same services rendered by larger units. Economies of scale are undeniably present in the work of government as they are in the private sector.

But the possibilities of increasing efficiency by increasing the size of governmental units can easily be exaggerated. For one thing, economies of scale are not inexhaustible. This point can be illustrated by a municipal tennis facility. If it is operated by a full-time caretaker, the cost per hour of play will be lower if the facility contains 4 courts than if it has only 1, and lower still if it contains 12 courts. But if that one caretaker can deal adequately with only 12 courts at the most, then any greater number of courts will require at least two full-time caretakers. So the cost per hour of play will be higher for an installation with 16 courts than for one with 9; and 24 courts will be no cheaper than 12. Moreover, as the number of courts and of caretakers is multiplied, a scale will be reached at which new costs of coordination and management are encountered, such that the cost per hour of play will be higher on an installation of, say, 120 courts than on one with 12 or 24. It follows that a town large enough to keep 12 tennis courts busy could provide that service at least as cheaply as could a town with ten times as many people. The complications of this hypothetical example are borne out by empirical studies of economies of scale, which show that, of the units producing any good or service at a given moment, the very largest are sometimes less efficient than those not quite so large and the very smallest are less efficient than those not quite so small, and, further, that over quite a large medium range, size does not affect efficiency. It follows that the largest possible unit of government—national or, if it came to that, international—would not necessarily be the most efficient provider of municipal services.

A second reason why the principle of efficiency does not imply *unlimited* enlargement of governmental units is that the most efficient scale of output differs for each kind of public service. It is intuitively obvious, for instance, that telephone services are more efficient if nationally or even internationally organized, whereas ordinary public libraries, say, will be less efficient if national rather than local in scope—the difference arising in some measure from the fact that books cost far more to transport than do telephone messages. It is similarly obvious that military defense is more efficient if operated as a service of the central government, whereas fire protection is more efficient if operated as a local service—or at least it is no less efficient that way than if operated as a centralized national service.

From the standpoint of efficiency alone, then, the ideal would be realized if every governmental service were rendered by a unit large enough to achieve the greatest possible economy of scale; and, since the most efficient scale is different for different services, each service would be rendered by a separate unit of government. National government would be charged with providing many services, inasmuch as these services could be operated most economically on a scale as large as, or even larger than, the extent or population of the whole country. The outcome of this ideal solution would be a multiplicity of autonomous units of government, many overlapping in their geographical domains. This result is not unthinkable; such a concept may help explain the existence in American local government of overlapping county, city, water district, park district, transportation, and other authorities. But it is not an altogether happy outcome. Citizens cannot exercise real or direct control over public authorities when there are so many that the ballot becomes interminable. And, worse, actions of autonomous public agencies whose domains overlap are not self-contained but influence each other and sometimes conflict with each other, thus creating problems of coordination that may be insoluble as long as the units retain full autonomy. Such overlapping results in "spillover," which arises whenever the actions of one autonomous unit of government affect members of any other autonomous unit of government. Overlapping, and the resulting prevalence of spillover, means that if each government function were assigned to an autonomous unit of the ideally efficient size, no government function would be performed as efficiently as it could be if the actions of all autonomous units were properly coordinated.

To summarize the dictates of the principle of efficiency, units of government should be large enough to reach the most efficient scale for producing the services they are required to render; this means that they should be rather large, often national, or even supranational, in scope, and of as many different sizes as would be required by the particular economics of each separate service. It might be added that since the economics of each service is more or less frequently altered, especially by changes in technology and also by changes in the relative cost of labor and capital, the scope of each unit of government will have to be altered more or less frequently in the effort to keep it at the most efficient scale.

Reconciling the Principles

Self-determination militates against large units of government, efficiency against small units: that is the fundamental dilemma faced by all governments that are democratic in the English liberal sense.

Ordinarily, when rational men face such dilemmas they overcome them by compromise. If cheap food tastes bad and delicious food costs too much, reasonable men follow a middle way; and, indeed, the middle way in all things has been identified since Aristotle with the road of reason. The reasonable solution, then, to the dilemma of local government might be to make all units middle-sized, to have one single tier of local government within which all units were large enough to capture some economies of scale and small enough to allow for a degree of intimate democratic control (that is, a control that might be exercised by temporary *ad hoc* pressure groups rather than always through local branches of strongly centralized national parties). Middle-sized units would, of course, be subject to the criticism that they were too big for comfort and too small for efficiency. But this criticism could be shrugged off as an unsophisticated refusal to recognize the conditions under which men cannot avoid living. Alternatively, the criticism might be answered by a piece of economic analysis demonstrating that when those two opposed constraints are operating, the optimal solution is reached at the size where the marginal value of efficiency just equals the marginal value of democratic control.

The constitutional form that would emerge from these considerations would be a central government—to do those things that only a central government can do (such as coercively adjusting conflicts between local authorities when all efforts at voluntary compromise have failed)—and one tier of local governments, all of about the same size, a size (being measured by relevant indices of population, income, and area) determined by the optimizing criterion mentioned above. This pattern would be entirely compatible with localized administration: central government could maintain sub-post offices in every neighborhood and local governments could maintain schools or sub-town halls in whatever profusion was administratively efficient. But units of government (in which broad policies are determined rather than administered) would exist at only two levels: that is, any citizen would vote for only two slates of representatives, one for his national government and another for his one and only local government.

A very different way of reconciling the two conflicting principles would be to try to satisfy each by establishing successive tiers of local government. Some would be very small, in order to satisfy the desire for self-determination, others large, to achieve maxima of efficiency in certain public services. This solution implies a great number of overlapping authorities of different sizes, all relatively large, each adjusted so as to permit the most efficient performance of the particular service with which it is charged.

Such a compromise would be full of imperfections. The very small units would perform whatever services they rendered very inefficiently. The largest units would not allow for any close approximation to unanimously approved policies or to direct democratic control, but would tend toward bureaucratic control and toward policies satisfying only vociferous or powerful minorities of the governed. Every voter would be called on to elect a large number and variety of representatives from among far more candidates than he could possibly know or even care much about; in consequence, he would tend to become bewildered, apathetic, or a willing accomplice to an oligarchy of party officials, to whom he would give his proxy by voting a straight ticket on most ballots, if not on all. So, although this complex compromise would establish units that were potentially efficient in that they were of appropriately large scales, they would not necessarily operate at anything like their potential efficiency, for want of adequate surveillance by an informed electorate. And, of course, the profusion of geographically overlapping autonomous authorities would give rise to manifold problems of spillover, which it might strain the capacity of central government to resolve. The task of development planning, for instance, becomes impossibly complex if the plan for any given area must take into account unpredictable and uncontrollable actions of a great many autonomous units of government that control some aspect of life in that area.

The present structures of local government in both England and the United States are much closer to the second, or complex, kind of compromise than to the first. This is because they have evolved in response to a variety of different intentions imposed during many centuries and in response to the influence of endless historical accidents. Both systems are workable and, when compared with the standards of most places at most times, eminently satisfactory. Both, however, exhibit ample defects, and it may well be argued that reform ought to aim toward reorganizing the structure rather than perpetuating or exacerbating existing complexity. How far one should or can go in the direction of simplicity at any given moment is, of course, a delicate political question. Simplification not only disturbs the present functioning of institutions and antagonizes present officials whose positions are to be abolished, but it also upsets the public, many of whom have learned, by considerable effort, to live with the present system and do not relish learning how to cope with altered arrangements, even if those promise to work better in the long run. But if the faults in the existing system are really bad enough to justify the disruption that comes with reform, then the direction in which reform should move is clearly toward simplicity and regularity of structure.

The Redcliffe-Maud Report: An Exercise in the Logic of Political Organization

The most painstaking of many official inquiries into English local government was carried out by a Royal Commission under the chairmanship of Lord Redcliffe-Maud during the period 1966–1969. In its 1969 report,[1] the Commission concluded that the current system of local government needed to be drastically simplified because it was too fragmented, uncoordinated, and inefficient.

[1] Royal Commission on Local Government in England, 1966–1969, *Report*, vol. 1, Cmnd. 4040 (HMSO, 1969). All text references unless otherwise indicated are to this report.

British Local Government in 1969

The principal components of that system of local government were arranged in two tiers, the main powers held and the main revenues raised by the counties and county boroughs. All of England was divided into 45 "administrative counties." Within their geographical boundaries, but entirely independent of their authority, were scattered 79 "county boroughs," cities ranging, with few exceptions, from 50,000 to 500,000 in population. Together the counties and county boroughs made up the first-tier authorities, which thus numbered 124. (The Greater London Council is another first-tier authority, but, since London presents special problems, is differently organized, and is subject to special legislation, it was omitted from the Commission's inquiry.) The second tier consisted of "districts" into which all counties (but not the county boroughs) were subdivided. Of these, there were three species: (1) "non-county boroughs" or "municipal boroughs," towns ranging in population from 2,000 to 100,000 (and averaging 30,000), of which there were 227; (2) "urban districts," somewhat smaller towns, ranging from 2,000 to 120,000 (and averaging 18,000), of which there were 449; and (3) "rural districts," with populations ranging from 1,500 to 86,000 (average 24,000), of which there were 410. Each of the 124 first-tier units and 1,086 second-tier units was governed by an elected council. In total, therefore, 1,210 self-governing units provided the local government for some 45 million Englishmen—not, perhaps, by American standards, an extravagant supply. In addition, rural districts were subdivided into a very large number of parishes, which enjoyed a minute power of self-government (para. 373).

The sheer number of local authorities was not, however, what mainly impressed the Commission with the need for reform; rather it was the finding that power and responsibility were too fragmented among the various authorities.

On the surface, the division of functions between the various units of local government did not look particularly fragmented. First of all, there was no fragmentation at all within the 79 county boroughs (cities), since they contained no self-governing districts. Second, the division of functions between the 45 counties and their 1,086 constituent districts seemed on paper to be reasonably sharp. Counties were responsible for planning, roads, education, health and welfare, police and fire protection, provision of parks, and several lesser services. The districts were responsible for public health—including water and sewage, garbage collection, and street cleaning—and for housing, which included providing public housing, administering subsidies to private housing, managing slum clearance, and enforcing certain building regulations.[2]

---

[2] A detailed outline of functions is given in Annex 3 to the *Report*, pp. 322–330.

Fragmentation resulted from this clear underlying scheme largely because the counties were required or allowed to "delegate" many of their functions to district councils. For example, any district with a population over 60,000–(in 1968, 55 of the 1,086 districts had reached this size [p. 333]–could demand the right to wield delegated planning powers, which would enable them to exercise and enforce planning control (though not to prepare the plans, the latter being a responsibility that the county councils might not delegate). Planning powers were also required to be delegated to such other district councils as the Minister of Housing and Local Government decided to designate, and could be delegated to still others if the county, district, and Minister agreed (p. 322). Responsibility for education was similarly split. Districts with over 60,000 population (but not the 20 rural districts of that size) could claim the right to exercise delegated powers to control primary and secondary education, and all other districts could be authorized to do so if they were able to persuade the Secretary of Education and Science that special circumstances existed (p. 324). On the other hand, though counties were required to delegate certain health functions to larger districts desirous of performing them, they could not under any circumstances delegate the responsibility to provide ambulance services (p. 325). And, finally, to consider a minor but illuminating example, counties were responsible for dealing with diseases of animals–and so, to the exclusion of the county's power, were noncounty boroughs that had had a population of at least 10,000 in 1881 (p. 327). In many instances of delegation, so the Commission affirmed, the two authorities–the county retaining the titular power and the district the actual–maintained duplicate staffs and disagreed as to policies and practices (paras. 150–155). The Commission concluded that "true delegation of power from one local authority to another is hardly possible, since both authorities remain responsible to their own electorates" (para. 70; cf. paras. 89–92).

Another form of fragmentation arose because the counties contained within their geographical limits the county boroughs, that is, the large autonomous cities. Lancashire, the extreme instance, contained as enclaves some dozen county boroughs, whose total population was over half as large as the county's and whose taxable property was of the same order of magnitude. In their power to make land-use plans, the county and the county boroughs that it surrounded were equal and independent; and it needs no tedious demonstration to show how difficult or ineffective land-use planning must be under such circumstances, where the decisions of each planning authority may be inadvertently, if not deliberately, countermanded by the decisions of many neighboring authorities (para. 87–89).

Because of this fragmentation of responsibility, the central government had intervened to solve certain problems. It had done so "by producing

regional plans itself, by the appointment of regional economic planning councils, by persuading local authorities to work together on land-use and transportation surveys and on sub-regional plans, by taking power to establish
passenger transport authorities." But, concluded the Commission, "none of
these devices is satisfactory, since none puts responsibility squarely on local
government or provides for continuous and comprehensive planning allied
with power to implement the plans" (para. 87). In order to make local
government viable, the Commission argued, it must be reorganized so as to
eliminate fragmentation, and this meant that the number of autonomous
units had to be drastically reduced. The institutional structure and the map of
local government needed to be recast, and the great question the Commission
faced at this stage of its reasoning therefore was: how large should the units
of local government be? Another way to pose the same question is: how
should the terrain of England be subdivided for purposes of local government?

The Commission's Merger of City and Countryside

A first step toward answering the question of scale—especially when presented in its geographical aspect—was the Commission's decision to abolish
specifically urban and rural units of local government. Urban versus rural is a
distinction deeply embedded in the structure of English local government;
moreover, it seems to rest on great and continuing differences between the
ways men live in town and in country, which ought to be reflected in the
services that government provides for countrymen on the one hand and town-
dwellers on the other. But the Commission justified its proposal to amalgamate the local government of town and country by the argument that they
are interdependent.

> Town and country have always been, and must be, interdependent. . . .
> The enormous increase in personal mobility . . . has vastly increased the
> interdependence; and a local government structure which does not recog-
> nise this does not correspond with the realities of life. People from the
> countryside come into the towns for shopping . . . ; people who work in
> the towns increasingly live out in the country. . . . Moreover, a great deal
> of the building needed now by the people living in overcrowded towns . . .
> will have to take place . . . in areas now rural or semi-rural.   [para. 86]

That town and country in England and everywhere else are interdependent,
nobody can doubt; but so is every part of any nation interdependent more or
less with every other part; and if interdependence necessarily implied that the
geographical regions ought to be under a common government, then the only
acceptable level of government would be national and central. If the case for
local government depended on an area's social and economic isolation from

all others, then local government could never be justified. The Commission's decision on this score, in other words, flatly contradicted the principle of self-determination, if one holds—as it seems hard not to—that country-dwellers and town-dwellers, today in England, still differ fairly systematically in attitude toward questions of local public policy. An example of such systematic disagreement that comes readily to mind bears on the supply and maintenance of local access roads in rural districts. Country-dwellers obviously want more of them than do city-dwellers; a unit of local government that incorporates both town and country would tend, if decisions were made by referendum, to supply many such roads when farmers were in a majority and few when town-dwellers were. If the two groups had each its own local government, the rural population could have as many as it was prepared to finance, while the urban population would have neither the responsibility nor the power to decide anything about rural access roads. The difference between the tastes of those who live in town and those who live in the country is, of course, a matter of fact, and, while the Commission may have been correct in judging that they are not different enough to matter, it is not a judgment whose accuracy is absolutely obvious.

What is implied for local government by the unquestionable fact that town and country are related and the second unquestionable fact that they disagree? The latter suggests separate governments for town and country, but does the former suggest common government? A case for it might be made along the following line of reasoning, which underlies the endorsement of "balanced communities." Suppose that two adjoining neighborhoods are inhabited, the one exclusively by rich people, the other exclusively by poor. If each had a separate local government, each would supply a different mix and level of public services according to its tastes and resources. Differences might arise that would offend against the standards of the nation: if, for instance, bright children in the poor community went without schooling while stupid rich children had better education than they could digest. This inequality could be abated, and the independence of the two distinct local governments maintained, if the national government were to redistribute income between the two. Redistribution *might* also be accomplished indirectly by merging the two local governments into one, but then again it might not, depending on whether the homogeneous rich or the homogeneous poor community outweighed the other in deciding the government's policy, and depending on their views of what that policy should be. The advantage that *might* accompany this indirect and uncertain method of redistribution by merger of local governments is this: as long as they were each separately ruled, the two neighborhoods might regard each other as strange and hostile; conjoined in a single unit, they might come to regard each other with the tolerance that men extend to members of their "own" group. The premise is that common government makes good friends. It is not, of course, consistent with the premise that strong fences make good neighbors, or with the history of chronic separatism in multinational states. Still, it may be as much true as false; and it is enough to say here that the Commission did not closely examine the disadvantages of merging the local government of town and country.

Efficiency and the Scale of Local Government

Once the Commission decided in this way to regard England (except for four metropolitan areas) as homogeneous for purposes of local government, it needed only to cut it up into a convenient number of pieces. It approached the task by trying to establish lower and upper limits to the population of a unit. The principle of efficiency was invoked to determine the lowest population compatible with adequate and economical public service. The Commission looked hard at economies of scale and took advice from many experts. These said that the smallest efficient sizes were: for police, a unit with population of 500,000; for education (other than university), 200,000 to over 500,000; for welfare, 50,000 to 100,000 (paras. 94, 110-144). The Commission concluded that the overall minimum should be 250,000 (paras. 256-265). As to the maximum, various experts suggested that no diseconomies of scale would be encountered until populations exceeded 1 million or even 3 million, if then. But the Commission decided that the general management of a local government, as distinct from the provision of any specific public service, does tend to lose efficiency as its population approaches the millions. And it further supported this conclusion by referring to self-determination—"democratic control" and the "citizen's real sense of belonging" (paras. 266-276). For both reasons, it settled on a maximum of about 1 million.

Implicit in these extremes of 250,000 and 1 million is an average of about 600,000, which in turn implies about 60 units of local government to serve the 40 million Englishmen who live outside Greater London. In the event, the Commission mapped out 61 areas, ranging in population from the 3 million of the Manchester metropolitan area to the 195,000 of the Halifax area, and of which 6 exceeded the general maximum and 5 fell below the minimum (pp. 311-312). All but the 3 largest of these 61 areas would be "unitary," that is, would contain no self-governing subdivisions. The three largest (the metropolitan areas of Manchester, Birmingham, and Liverpool) would, on the other hand, be subdivided into metropolitan districts, of which there would be 20, containing, on the average, 400,000 people (p. 313). As between the 3 metropolitan areas and their 20 constituent metropolitan districts, functions would be divided so that the former would provide "environmental services"—planning, transportation, and major development—while the latter would provide "personal services"—education, health, social services, and the like. The drawing of the lines on the map of England, as disputed by several members of the Commission,[3] was based on a desire for physiographical and social coherence within each area.

Nevertheless, having decided that, for reasons of efficiency and self-determination, the structure should consist of 81 "main authorities" (58 unitary areas, 3 metropolitan areas, and their 20 metropolitan districts), the Commission was not yet satisfied. It felt that two additional tiers were essential to meet the residual needs for efficiency, on the one hand, and for democracy, on the other. They accordingly designed a second tier within the 58 unitary areas: "local councils," whose boundaries were none other than the existing boundaries of the county boroughs, municipal boroughs, urban districts, and parishes, and whose number amounted to something between one thousand and several thousand. It might seem that by reintroducing this second tier, comprising practically all the units in the previous second and third tiers, the Commission would have undone all its effort to simplify and rationalize. This possible objection was met by the decision, not altogether unambiguous, that the local councils would not really be governments, but organized pressure groups designed to force their views on the attention of the "main authorities," who would have all the powers of governing. Local councils would be principally, as the Commission put it, "institutions for local self-expression" (para. 368), which the main authorities would be obliged to consult about any decision affecting its special interests. In addition, local councils would have powers, though no duties, to provide public services supplementary to those provided by the main authorities insofar as they were prepared to pay for them (paras. 383–392). But they would have no control over the main authorities nor any mandatory duties, which sufficiently underlines that their nature would be advisory rather than sovereign. Would participation in this manner satisfy men's urges toward self-determination? It seems a pale substitute for the real thing, but, as it is grafted onto a system that gives the citizen a right to participate more actively in electing his representative to national and local governments, it might be enough.

Not only did the Commission conclude that the main authorities were too large for democratic comfort; it also concluded that they were too small for efficiency in one respect—local physical planning as it relates to national economic policy. The Commission accordingly proposed that the 61 areas and Greater London be grouped in 8 "provinces," each controlled by a council elected from its constitutent areas with an admixture of about 25 percent of coopted councillors. The function of each council would be to make a

---

[3] See the *Memorandum of Dissent*, vol. 2 of the *Report*.

"strategic plan" for future development, settling "the framework and order of priorities within which unitary and metropolitan authorities will work out their own planning policies and major investment programmes," and drawn up "in the closest collaboration with the unitary and metropolitan authorities and with central government" (para. 412).

To sum up, then, the structure proposed by the Commission consisted of three tiers: (1) a level (or, in the three metropolitan areas, two levels) of local government; (2) a level below to organize the expression of local sentiments; and (3) a level above to coordinate planning (para. 284). It should be emphasized that only the main authorities would be empowered to *govern*, while the levels above and below would be empowered to perform a spiritual and a technical function. The proposed change in structure is therefore quite radical: 1,210 existing units of self-government would be compressed into 81. No doubt fragmentation would be reduced and efficiency might be increased, but it is harder to forecast whether the change would create either the sense or the reality that local powers of self-determination had been strengthened.

Since the Commission published its report, a Conservative government has been elected and has found so sharp a centralization of local government uncongenial to its principles or tastes. Few of the existing local governments were eager to be dissolved; and, since most local governments were controlled by Conservative councillors, their reluctance carried special weight with a Conservative central government. Indeed, the reforms pushed by the government tend much more to perpetuate the present system of two effective tiers of local government. Redcliffe-Maud himself subsequently endorsed a compromise between the Commission's scheme and some two-tier scheme as "both necessary and practicable."[4] This concession represented a reliable estimate of political possibilities at that point in the evolution of British governmental institutions.

Efficiency in the Management of Local Government

So far the question of efficiency has been considered only from the standpoint of economies of scale. But obviously the realized, as distinct from the potential, efficiency of government, depends also on the quality of its internal organization and operation. And all the misgivings during the 1960s about the structural format of English local government were matched by doubts about its internal machinery. The Redcliffe-Maud Commission accordingly considered the matter of management, basing itself largely on the work of an earlier Committee on the Management of Local Government, which had also been chaired by John Maud (later Lord Redcliffe-Maud), and was therefore

[4] *The Times* (London), August 7, 1970, p. 7.

known as the Maud Committee, a similarity that has not failed to produce confusion.[5]

In management of government, as of any other organization, two separate functions can be distinguished. Policy decisions must be made to determine what ends the government will pursue for the time being. Technical decisions must then be made to determine how the means available can be directed most efficiently toward the accomplishment of the chosen goals. This distinction of functions marks the line that separates politics from administration, and the line that, in a democratic order, separates elected executives and legislators from civil servants and public employees. Within a democracy all units of *government* must contain elected officials who make policy, though every state contains units of *administration* that may be staffed entirely by appointed employees. Units of administration are assigned their goals by governments, but every unit of government (properly so called) determines its own policies, and democratic governments have their policies set (however indirectly) by the governed. All this is familiar ground, as is the difficulty of distinguishing with absolute clarity between a policy decision and a technical decision.

From this distinction between political and administrative functions arises one of the chief problems of management to which the Maud Committee addressed itself. All existing units of local government were governed by elected councils, and each of the separate functions of the local governments, such as housing or education, had come to be governed by a separate specialist committee of the council, which reported directly to the whole council. So strong was the tradition of the amateur in English government—or, what is equivalent, so strong the tradition of self-government at the local level—that, on the one hand, councillors devoted large amounts of time to carrying out administrative (as well as political) duties connected with their specialist committees, while, on the other hand, they served without pay. The division of labor between councillors and salaried officials therefore became blurred: committees of the council, going beyond their political responsibilities to determine public policies, strayed far into the administrative domain of civil servants. The Maud Committee, and, following it, the Redcliffe-Maud Commission, concluded that administration would be more efficient if the executive functions performed by councillors were left strictly to their paid professional officials. The terms in which the Maud Committee put this point are revealing:

---

[5] Ministry of Housing and Local Government, Committee on the Management of Local Government, *Report* (HMSO, 1967); henceforth cited as Maud Committee, *Report*.

Reference is made [in an earlier section] to the unfortunate effects that a rather narrow interpretation sometimes placed on the word "democracy" has had in local government in this country. It is thought that unless members [of a council] determine how the smallest things are to be done, they are failing in their duties, and that to allow any but the most trivial discretion to an officer is undemocratic. The effect of this is to force issues however trivial upwards to the top for consideration in the committees. This in turn involves principal officers and their immediate subordinates in dealing with matters on their way to the committees which would otherwise be disposed of at a lower level. It is perhaps symptomatic of this tendency for issues to be dealt with at the highest level that letters are signed by heads of departments or by their subordinates writing the head of department's name. *We recommend that local authorities adopt the guiding principle that issues are dealt with at the lowest level consistent with the nature of the problem.* [Maud Committee, *Report*, vol. 1, para. 152]

The demands of democracy can certainly be exaggerated, but so can those of efficiency, and the recipe that all matters should be decided at the "lowest" possible level of administration is far too blunt a technique for weighing the conflicting needs to make speedy decisions and to make decisions that are acceptable to a majority of the governed, or at least to a majority of their elected representatives.

The second chief problem noted by the Maud Committee was the difficulty of coordinating the work of the various service committees of a council. A large city might have over 30 service committees and nearly 100 subcommittees in its council (para. 479). Each of the committees reported directly to the whole council: there was no other formal apparatus for coordinating their work, though in some instances coordination was facilitated by the clerk to the council, then the chief officer of most local governments. Councils were apt to be large: the largest had over 150 members; county councils averaged 86; and even the district councils averaged about 25 (Maud Committee, *Report*, para. 17). To ask a council of this size to coordinate the work of dozens of service committees, each of which was said to be tempted "to cling to their preoccupation with details and with supervision of routine" (para. 480), is clearly to ask a good deal.

What is the best method, in the abstract, for coordinating the actions of separate parts of any organization? What is the maximum or optimum capacity of any group to coordinate the activities of its constituent parts? In the abstract, the question does not seem to permit any simple or exact answer. A multiplicity of distinct functions is carried out within any organization: that fact is half the essential definition of an organization. The well-known efficiencies that result from specialization of function tend always to increase the number of distinctive duties within an organization of any given size. With each such increase, the task of coordination increases far more rapidly. If two distinct departments require one message to coordinate their decisions, three departments require three messages, four require six messages, and ten require forty-five. If we assume for the sake of illustration that each department coordinates each action with every other department, the *least* number of messages required is $\frac{N^2-N}{2}$, where N represents the number of departments. If coordination were turned over to a special agency, it would require at least $2N$ messages to collect information from each department and then to command each one to make the necessary adjustments. But all of this radically underestimates the true burden of coordination, especially when it involves persuasion and complex compromises. No trick of reorganization can radically reduce the burden of coordination, since that burden is fundamentally determined by the number of distinctive functions that must be coordinated, not by the number of separate departments. Whether two distinctive functions are performed by two separate departments or within a single department does not fundamentally alter the labor of coordinating them—except insofar as boundaries within a department are easier to cross than those between departments—but alters only the place where coordination is done. If the case were otherwise, all problems of coordination could be eliminated by abolishing all the departmental subdivisions within an organization. The continued insistence of all sizable organizations on erecting departmental subdivisions proves, on the contrary, that complete centralization is no more universal a solution to the problem of coordination than would be complete decentralization. This is not to say that any, or every, particular internal structure is just as good for a particular organization at a given time. Performance may be improved by adjusting the structure of an organization; but unfortunately there is no scientific method for determining in advance the most efficient departmental structure for any organization. Nor can any structural scheme reduce the costs of coordination below a minimum level that is dictated by the number of distinctive functions to be performed, the probability that departments will clash with each other, and similar aspects of the organization's makeup. Any general scheme to improve coordination is therefore prima facie suspect.

To cure the management ills of English local government, the Redcliffe-Maud Commission suggested that each of the new local governments should establish a "central committee, board or body of some kind, by whatever name it may be called" (para. 486) to carry out the chief work of coordination. It would be a committee of the council, standing between the council and its various service committees, which the Commission recommended should be reduced to a much smaller number. Just how the central committee would work, the Commission left to each local authority to decide for itself, while adding this proviso: "It is, however, a radical change from traditional practice that we seek. The central committee must be at the core of the administration; and the proliferation of committees must be ended" (para. 495).

These two proposals—to leave more of administration to professional managers and to leave more of policy making to a small central committee—might in fact improve the working of the machinery. They might also reduce the directness with which a citizen can, by prodding his elected representative, alter the effective decisions of his local government. A government made more efficient may be a government made less responsive to the wishes of the citizen, though this offsetting disadvantage need not be inevitable. Still, the possibility is live enough so that proposals for improved administration of government should not be received in the faith that such improvements are all gain and no loss.

Applicability to the United States

Could the proposals of the Redcliffe-Maud Commission be usefully imported into the United States? The first way to answer this question is to assume that "the urban problem"—or, more exactly, that part of it which is perpetuated or aggravated by imperfections in the structure of local government—is fundamentally the same in the United States as in England. If so, the remedies proposed for the English case should apply in some fundamental degree to the American case, after proper allowance has been made for differences of detail between the circumstances of the two cases. The applicability of each of the Commission's main proposals is considered, in this spirit, below:

(1) *Amalgamate town and country for purposes of local government.* This is already done in the United States by the states and the counties, though not always in a manner that satisfies both town and country. But to eliminate all lower tiers of local government would practically abolish *local* government: because of the size of the country, the distance of most citizens from the seats of state and even county governments would be too large to create a vivid sense that those who govern live in the "same place" as the governed.

It is true enough that in America, as in England, town and country are now harder to distinguish than they used to be; consequently, land use should not be planned on the assumption that the land used by members of a large urban settlement is or will continue to be confined within the political boundaries of the local planning authorities. In order that land-use planning should deal with sufficiently extensive areas, it might be useful to require that local plans conform to development plans for the state in which they lie. This still leaves the problem of places, such as the megalopolis Boston-Washington, which spread over several states. One possible device might be a regional planning province for the Northeast, established under federal statute, governed by a council elected on some pro rata basis by each of the nine or ten states involved, its decisions to have binding force on the development plans of each participating state and thereby, of course, on each unit of local government within the megalopolis.

(2) *Establish regional planning provinces.* The population and areas of American states are generally large enough to make each an adequately extensive planning area. Of course, spillovers cannot fail to occur across state borders, but spillovers cannot be avoided by anything short of exclusively national government. As things stand now, little would be gained by grouping the states into six or eight regions, although the megalopolitan areas may need special treatment as suggested above.

(3) *Establish small local units with a right to be consulted by the operative tiers of local government.* The right to be consulted may create a comforting sense of "participation." It may ward off the displeasure of living by rules that one dislikes. It may even, at best, ease the transmission of the citizen's sentiments to his rulers. But if it created an illusion of power that was subsequently dispelled by the realization that being consulted does not mean getting one's way, it might end by having done more harm than good. Subject to these cautions, it might be worth experimenting with officially established neighborhood councils enjoying a statutory right to be consulted by local government in matters affecting the particular neighborhood, and perhaps in all matters of broad public policy.

(4) *Simplify the structure of local government.* A going system of local government ought not to be tinkered with light-heartedly or even optimistically. Particular diseases of detail should be doctored with medicines mild enough not to upset the whole system. Drastic reform can be justified only if the existing system is drastically out of order *and* if the disorder stems from inadequacies in the structure and not from causes of a quite different kind. To the extent that the American "urban problem" arises from difficulties in race relations, family relations, and the like, it is not likely that it will be solved by reorganizing local government. Reorganization is so far from a panacea that it cannot even be guaranteed to cure difficulties that arise from faulty organization.

Translated more or less literally to the American scene, the Redcliffe-Maud Commission's proposals would mean abolishing all local governments in the United States other than the 50 states. This would mean that many Americans would find themselves living 200 miles or more from the seat of their "local" government, which seems excessive.

Multiplicity of overlapping local authorities creates problems that have been mentioned earlier. Many such authorities provide public services that might be and have been furnished by private enterpreneurs, such as water companies and bus lines. Such functions, whoever performs them, are commercial rather than governmental: they need involve no element of coercion and, unlike such services as national defense, need not depend on public financing (even if they are thought to need public subsidies). If such authorities were converted into corporations (whether private, public, or mixed) whose directors were appointed by the local governments concerned, the change might do away with the illusion that voters can effectively control those bodies by electing their directors, and might replace it with a more effective control by local government.

(5) *Increase professional management of local government.* To the extent that local government is engaged in governing as distinct from providing quasi-commercial services, a distinct limit is set to the amount of professionalism that is compatible with democratic control. Officials with life tenure do not follow the election returns: on the contrary, they are given security of tenure precisely to insulate them from the electorate. For this reason, the idea of nonpartisan *government* is at best a delusion and at worst a fraud. American local governments have moved much further on the road to professionalism than have the English, possibly far enough, on the average. No doubt some or many American local governments could be improved by introducing more professionals into various functions that are not now performed at all or well. But the great current problems of American cities are not such that they would melt away under the skilled hands of the professional civil servant or of any known sort of expert. The problems are there because of the very real and intense conflict among people who live side by side. Such problems are political and can only be resolved by the practice of politics.

To summarize, then, the suggestions that would emerge from a narrow and fairly literal transposition of the Redcliffe-Maud proposals to the American scene are:

(1) Plans of each local government should be required to conform to a strategic plan drawn up by its state.

(2) Plans of each metropolitan area that extends into two or more states should be made by a "regional" authority—wide enough to comprehend the metropolitan area and the several states comprised in the region.

(3) "Neighborhood" councils should be created which would enjoy a statutory right to be consulted by their local government on all matters affecting the neighborhood.

(4) Some of the purely service functions now performed by specialized and autonomous units of local government should be transferred to

quasi-public corporations whose officers would not be directly elected by the public but appointed by and supervised by elected officials of local government.

These may be useful hints, though they fall far short of constituting either a radical program to reorganize American local government or, for that matter, a particularly hopeful strategy for solving the problems of American metropolitan development. But perhaps one should expect no more from such an exercise, since to hope that one nation's problems will be solved by another nation's reforms may be like hoping that this man's rheumatism will be cured by the medicine that eased another man's gout.

A rather different and better way to take advantage of the Redcliffe-Maud proposals would be to recognize at the outset one basic difference between the conditions of local government in England and the United States—namely, the difference produced by the federal Constitution. The relative autonomy of state and local governments that the federal system preserves is not perfectly adapted to dealing with a nation whose urban population and urbanized areas are growing at a rapid pace.

The metropolitan areas of the United States spill over the urban governments, and the urban problem spills over the metropolitan areas. Housing used by some members of the Boston metropolitan area is located twenty miles or more outside the boundaries of the City of Boston, just as factories that employ residents of those outlying suburbs are located miles outside the boundaries of either the central city or the suburb. A broader example of spillover is furnished by recipients of welfare in the City of New York, many of whom have recently immigrated from other less wealthy states, just as many families who earn their livings in the City of New York now travel along highways of New Jersey in order to reach their residences in Pennsylvania. In order to keep such spillovers, as well as the continuing rapid growth of metropolitan areas, under a minimal degree of rational control, it is essential that metropolitan government be roughly coterminous with the real life of metropolitan areas. But this is exactly what the multiplicity and autonomy of American local governments make so difficult. When the problem of spillover was recognized in England, it could be met in part, albeit imperfectly, by giving the central government, under the Town and Country Planning Acts, powers to coordinate and even to censor planning by local authorities. In the United States no such solution is either constitutionally or politically feasible for the time being.

How could broad coordination of metropolitan land-use planning be achieved in a manner compatible with the continuing autonomy of state and local governments? One hint derived from the Redcliffe-Maud Report might be worth looking into. It is to establish a three-tier system in which, contrary to the usual expectation, the greatest degree of autonomy and power would be conferred on (or, rather, reserved to) the middle tier. Suppose that the fundamental and original power to make city plans were exercised by *the* government of a metropolitan area—that is, by a tier of government that hardly as yet exists. The tier below, consisting of the various units of government that now rule subsidiary parts of a metropolitan area, could be given a statutory right to be consulted by the main planning authorities, along the lines suggested by the Redcliffe-Maud Commission. And a tier above the main authorities, consisting of the state government or, in some instances, consortia of state governments, would have the function of coordinating the plans of the various main authorities within the state or the multistate region, though distinctly a function of coordinating by compromise and bargain rather than of command. Some such plan could preserve the main spirit of a federal system while introducing machinery better suited to dealing with the spillovers so characteristic of a highly urbanized and closely interwoven society. Fragmentation of local government, in the American system, is not only a vice but also a benefit; it is, however, an increasingly expensive benefit, and some move toward closer coordination of local policies (as distinct from outright integration of local governments) may be necessary to keep it from becoming too expensive to be afforded.

Library of Congress Cataloging in Publication Data

Main entry under title:

Reform as reorganization.

    (The Governance of metropolitan regions, no. 4)
    Includes bibliographical references.
    1. Metropolitan government—United States—Addresses, essays, lectures. 2. Metro-
politan government—Great Britain—Addresses, essays, lectures. I. Hanson, Royce.
II. Resources for the Future. III. Series.
JS422.R42   352′.008′0973   73-19348
ISBN 0-8018-1544-4